The Never-Ending Day

Marcus Chown is an awar... ...ormerly a radio astronomer at the C... ...Pasadena, he is now cosmology consultant of the weekly science magazine *New Scientist*.

Marcus's first popular science book, *Afterglow of Creation* – runner-up for the prestigious Rhône-Poulenc Prize – was published to much acclaim in 1994. *The Magic Furnace*, his second book, was published in Britain in 1999. In Japan it was chosen as one of the Books of the Year by *Asahi Shimbun*, the world's biggest newspaper and, in the UK, the *Daily Mail* called it 'a dizzy page-turner with all the narrative devices you'd expect to find in Harry Potter.' His third book, *The Universe Next Door*, was published in 2002. 'An exuberant book – a parallel universe where science is actually fun,' wrote the *Independent*.

Marcus lives in London with his wife, Karen, and their two children.

THE NEVER-ENDING DAYS
OF BEING DEAD

Dispatches from the Front Line of science

Marcus Chown

ff

faber and faber

First published in 2007
by Faber and Faber Limited
Bloomsbury House, 74–77 Great Russell Street
London WC1B 3DA

This paperback edition first published in 2007

Typeset in Bembo by Palindrome
Printed in the UK by CPI Bookmarque, Croydon

A CIP record for this book
is available from the British Library

ISBN 978–0–571–22056–4

Contents

CONTENTS

Acknowledgements

My thanks go to the following people who either helped me directly, inspired me or simply encouraged me during the writing of this book. Karen, Sara Menguc, Neil Belton, Henry Volans, Cara Defoe and Squid, Gregory Chaitin, Stephen Wolfram, Larry Schulman, Jim Hartle, Cris Calude, Paul Steinhardt, Neil Turok, Stephen Hsu, Anthony Zee, Frank Wilczek, Bernard Haisch, Alfonso Rueda, Frank Tipler, Max Tegmark, Victor Stenger, Andrei Linde, Paul Shuch, Seth Shostak, John Casti, Charles Bennett, George Johnson, Freeman Dyson, David Deutsch, Jim Crace, Pam Young, Brian Clegg, Patrick O'Halloran, Karen Gunnell, Jo Gunnell, Barbara Pell, Hazel Muir, Michael Brooks, Valerie Jamieson, Roy Perry, Sarah Topalian, Cliff Pickover, Sir Martin Rees, Alex Jones, David Hough, Fred Barnum, Stephen Hedges, Sue O'Malley, Spencer Bright, Ciaran Tomlin and all from the planet Treetops . . . It goes without saying, I hope, that none of these people are responsible for any errors.

To Geoff & Helen, Martin & Anne, and Shirley

Preface
Ultimate Questions, Ultimate Answers

We've come a long way. Once upon a time we thought the world rested on the back of a turtle and the Sun was a ball of molten iron 'not much bigger than Greece'. Now we have a theory of small things like atoms – quantum theory – which not only explains why the Sun shines and why the ground beneath our feet is solid but has also given us lasers, nuclear reactors and iPod nanos. In addition, we have a theory of big things like the Universe as a whole – Einstein's general theory of relativity – that predicts the existence of black holes and suggests that there was a beginning to time.

Nobody yet knows how the theory of the small meshes with the theory of the big – that's theorists' work-in-progress – but no matter. The point is that previous generations would have killed for the kind of knowledge we now possess about the world. It truly is a privilege to be alive today. For the first time in history, we have a good idea of the extent of the Universe – we can see all the way to the 'light horizon' that forms the boundary of observable space – and we have a good idea of the content of the Universe – we can count up the building blocks of the cosmos, 100 billion or so galaxies like our Milky Way.* And not only do we have an idea of the extent and content of the Universe but we also have a strong indication of how it came to be. The Universe burst into being about 13.7 billion years ago in a titanic explosion called the Big Bang and has been expanding and cooling ever since. Our Milky Way – along with all the other galaxies – simply congealed out of the cooling debris of the Big Bang fireball.

Admittedly, we still do not know exactly what the Big Bang was, what drove it or what happened before the Big Bang (or whether this is even a meaningful question). However, the remarkable thing is that we are the first generation with a realistic chance of answering such 'ultimate questions'. And not only these ultimate questions but a host of others, such as:

* What is beyond the edge of the Universe?
* Where does all the complexity we see around us come from?
* What are the limits of what we can 'know'?
* Is the human brain doing more than any computer?
* Where does the everyday world come from?
* Why do we experience a past, present and future?
* Why are loaded fridges difficult to budge?
* Will we ever find ET out in the Universe?
* Can life survive for ever in the Universe?

I address all these questions in this book. To answer them, I have talked to some of the most imaginative and daring scientists in the world. In discovering their extraordinary answers, you will learn – among other things – how the Big Bang might have been spawned by a collision

---

* The word billion is used in this book for a thousand million, 1,000,000,000.

between 'island universes'; how a single remarkable number contains the answer to every question we could ever ask; how the most widely accepted theory of the Universe's origin implies that Elvis is alive and well and living in another space domain (in fact, an infinite number of other space domains); how nobody can rule out the possibility that the stars are technological artefacts built by extraterrestrial intelligence; how a computer program a mere four lines long could be generating you, me and everything we see around us; how all of us might be resurrected in a computer simulation at the end of time.

The last possibility inspired the title of this book. According to one controversial physicist, when you die, you are fast-forwarded to the dying days of the Universe where you wake up inside the ultimate cyber reality. Stretching before you you will find a subjective eternity of existence – the never-ending days of being dead.[*]

The ultimate questions I tackle are by no means a definitive selection; they are simply ones that have intrigued me personally. Nevertheless, there are themes which tie many together. I have therefore grouped together the questions into convenient categories. First, there are those whose answers shed light on the 'nature of the universe' – questions such as 'What is beyond the edge of the Universe?' and 'Where does the Universe's complexity come from?'. Second, there are questions whose answers illuminate the 'nature of reality' – 'Where does the everyday world come from?' and 'Why do we experience a present?'. Finally, I address questions which address the place of life (and us) in the Universe – such as 'Will we ever find ET?' and 'Can life survive for ever in the Universe?'.

Some of the ultimate questions I address may at first sight seem

---

[*]   Thanks for that line to Jim Crace and his brilliant novel *Being Dead*.

abstract and esoteric. But it is a remarkable feature of such questions that they invariably have significance for our mundane everyday lives. There is no more basic question, after all, than 'Where does the everyday world come from?'. And, even the most esoteric question I ask – 'Does a single number contain the secret of the Universe?' turns out to have a bearing on the origin of human imagination and creativity and whether the brain is doing something more than any computer. I think this is the nature of science at the leading edge. It is ultimately about down-to-earth things we all care about – Where did we come from? Where did the Universe come from? What the hell are we doing here?

One last thing. The answers I present are not necessarily linked to each other. This is characteristic of science at the frontier, where new ideas are so new they have not yet been woven into the tapestry of accepted science. Some will stand the test of time and some will not. Some are even mutually exclusive. All the ultimate questions are hard questions. The hardest questions are always the most interesting ones. Answering them requires journeying to the very frontier of science – and, in fact, way beyond. Have fun!

Marcus Chown

Part One
THE NATURE
OF THE UNIVERSE

1
Elvis Lives

What is beyond the edge of the Universe? An infinity of other domains where all possible histories are played out

Whatever spot anyone may occupy, the universe stretches away from him just the same in all directions without limit.

Lucretius, 1st century BC

There are two things you should remember when dealing with parallel universes. One, they're not really parallel, and two, they're not really universes.

Douglas Adams, *The Hitchhiker's Guide to the Galaxy*

Far, far away, in a galaxy with a remarkable resemblance to the Milky Way, sits a star that looks remarkably like the Sun. And on the star's third planet, which looks remarkably like the Earth, lives someone who, for all the world, looks like your identical twin. Not only do they look the same as you but they are reading this exact same book – in fact, they are focused on this very line. Actually, it is weirder than this. A whole lot weirder. There is an infinite number of galaxies that look just like our own galaxy, containing an infinite number of versions of you whose lives, leading up until this moment, have been absolutely identical to yours.

If you think this is pure science fiction, think again. The existence of your doubles is no fantasy. It is an unavoidable consequence of the standard theory of our Universe. And it is no airy-fairy consequence either. If you could voyage far enough across the Universe, it is inevitable that you would run into one of your doubles. In fact, it is even possible to calculate how far you would have to go to meet your nearest doppelgänger. The answer is roughly $10^{10^{28}}$ metres.

In 'scientific notation', the number 10^{28} (10^{28}) is 1 followed by 28 zeroes, which is 10 billion billion billion. Consequently, $10^{10^{28}}$ is 1 followed by 10 billion billion billion zeroes. It is a tremendously big number. It corresponds to a distance enormously farther than the farthest limits probed by the world's biggest, most powerful telescopes. But do not get hung up on the size of this number. The point is not that your nearest double is at a mind-bogglingly great distance from the Earth. The point is that you have a double at all.

What you have just been let into is cosmology's embarrassing little secret. It is something cosmologists rarely like to mention in public. And who can honestly blame them?

But why does the standard theory of the Universe have such an extraordinary consequence? There are two reasons, it turns out. The first is 'quantum theory', our best description of the microscopic realm of atoms and their constituents. And the second is a popular theory of the first split-second of the Universe's existence, called 'inflation'.

Failures of the Standard Big Bang

Inflation is something which cosmologists have to bolt onto the standard picture of the Big Bang because, to put it bluntly, the standard picture does not work. It predicts things which are not what we see when we look out across the Universe.

According to the Big Bang picture, our Universe began in a dense, hot state about 13.7 billion years ago and has been expanding and cooling ever since. The main evidence for this comes from the galaxies – great islands of stars of which our Milky Way is one among at least 100 billion. They are flying apart from each other like pieces of cosmic shrapnel. The unavoidable conclusion is that they were closer together in the past. In fact, if the expansion of the Universe is imagined running backwards, like a movie in reverse, a moment is reached – about 13.7 billion years ago – when all of the Universe's matter was squeezed into the tiniest of tiny volumes. This was the moment of the Universe's birth – the Big Bang.

When anything is squeezed into a small volume – for instance, when air is squeezed in a bicycle pump – it gets hot. The Big Bang was therefore a 'hot' Big Bang. The evidence for this is in fact all around us today because the heat of the Big Bang was bottled up in the Universe and had nowhere else to go. Every pore of space is therefore still permeated by the 'afterglow' of the Big Bang fireball.[*] Because this 'heat radiation' has been greatly cooled by the expansion of the Universe over the past 13.7 billion years, it no longer glows as visible light. Instead, it appears as 'microwaves', a type of light invisible to the naked eye but which is familiar to us from radar, mobile phones and, of course, microwave ovens.

Tune your TV between the stations. About 1 per cent of the static, or 'snow', on your screen is due to this 'cosmic microwave background radiation'. Before it was intercepted by your TV aerial, the last time it interacted with matter was in the searing-hot fireball of the Big Bang. Taken together, the fact that the Universe is everywhere glowing with heat, and the fact that it is expanding, strongly supports the idea that, in the very remote past, all of Creation erupted out of a super-dense, super-

[*] See my book, *Afterglow of Creation* (University Science Books, Sausalito, California, 1996).

hot state. But, though this 'Big Bang' scenario explains so much, it cannot be the whole story. The reason is that it fails to predict what we observe in the Universe in three major ways.

For one thing, the standard Big Bang picture predicts that galaxies like our own Milky Way should not exist. Galaxies are believed to have arisen from regions of gas which in the fireball of the Big Bang were ever so slightly denser than their surroundings. This gave them slightly stronger gravity so that they were able to pull in more matter and grow more effectively than neighbouring regions. But this would have been a painfully slow process. The fully fledged galaxies we see about us today could not possibly have congealed out of the smeared-out matter of the fireball of the Big Bang in a period of time as short as 13.7 billion years. Cosmologists fix the problem by postulating the existence of a vast amount of invisible, or 'dark', matter. The extra gravity it provided would have speeded up galaxy formation by pulling together matter faster so that the galaxies could have formed in the available time.

Even with this fix, however, there is a second major thing which the basic Big Bang model predicts which we do not see. It predicts that the effect of every galaxy pulling with its gravity on every other galaxy should be to 'brake' the Big Bang-driven expansion of the Universe. Contrary to all expectations, however, physicists discovered in 1998 that the expansion of the Universe appears to be speeding up. Here the standard fix is to postulate the existence of 'dark energy', an invisible 'springy' stuff which fills all of space. Its repulsive gravity is said to be countering gravity and so remorselessly driving the galaxies apart.

But, even with the addition of dark energy and dark matter, there is a third thing that the Big Bang model predicts which we do not see. This is a slightly more esoteric matter but it has to do with the 'smoothness' across the sky of the cosmic background radiation.

How the Universe Got to Be All at the Same Temperature

If we imagine the movie of the Universe running backwards, eventually we come to the epoch when the Big Bang radiation originated – a period about 450,000 years after the moment of creation.[*] At that time, the observable Universe was about 18 million light years across.[†] This is unexpectedly large for such an early time. In fact, it is inexplicably large and poses a serious problem for the standard picture of the Big Bang. To understand why, it is necessary to imagine what happened to the Universe as it cooled in the immediate aftermath of the Big Bang.

Things never cool down evenly. So it is likely that parts of the rapidly expanding fireball cooled slightly faster than others. What normally happens in such circumstances – for instance, when a cup of coffee is left on a tabletop – is that any unevenness in temperature that develops gets ironed out. This is because heat continually flows from the hot regions to the cool regions, equalising the temperature.

There is a limit, however, to how fast heat can flow. It is set by the speed of light – the cosmic speed limit. Nothing can exceed it – and that includes heat. The speed limit has no bearing on the flow of heat in something as small as a cup of coffee. But it is hugely significant for

[*] Actually, the cosmic background radiation comes from an even earlier time than this. After all, it is the relic heat of the Big Bang fireball. Nevertheless, it was only at about 450,000 years after the birth of the Universe – at the so-called epoch of last scattering – that it broke away from matter and was able to fly freely across space. It has space. It has been flying freely across space ever since – the oldest fossil in Creation – carrying with it an imprint of the Universe close to the beginning of time.

[†] A light year is the distance light travels in a year. Since the speed of light in a vacuum is about 300,000 kilometres a second – fast enough to go from the Earth to the Moon in about 1¼ seconds – a light year is about 10 million million kilometres.

something as big as the Universe – even something as big as the shrunken primordial Universe.

Light, by definition, travels at a light year a year. This means that, when the Universe was 450,000 years old, light could have travelled no more than 450,000 light years. But, as pointed out, the Universe at the time was 18 million light years across. Light – and heat – could therefore have spanned only a few per cent of the diameter of the Universe.

It follows, therefore, that, if one side of the rapidly expanding fireball cooled down just a fraction faster than the other, it would have been impossible for heat to have flowed from the hotter side to the cooler side to equalise the temperature. There simply would not have been enough time since the beginning of the Universe.

The standard picture of the Big Bang therefore predicts that the 450,000-year-old Universe must have had an uneven temperature. Furthermore, the cosmic background radiation, because it was mixed in with the matter of the fireball and shared its temperature, should also have an uneven temperature. But this is not at all what astronomers see. When they look at the Big Bang radiation coming from widely separated parts of the Universe – which in practice means pointing a radio telescope in widely different directions in the sky – they see a remarkable constancy in its temperature. In fact, to within much less than 1 part in 10,000, the temperature of the cosmic background radiation is exactly the same in every direction in the sky – a chilly 2.726 degrees Celsius above absolute zero.[*]

The standard Big Bang picture tells us that heat could not have flowed back and forth across the early Universe and ironed out any temperature

---

[*] Absolute zero is the lowest temperature attainable. When an object is cooled, its atoms move more and more sluggishly. At absolute zero – -273.15 degrees Celsius – they stop moving altogether (apart from a residual jitter which is a consequence of quantum theory).

differences. But our observations say emphatically that it did.

There are several possible ways to resolve the conflict. All of them involve bolting something new onto the standard picture of the Big Bang.

One possibility is that, in the early Universe, the speed of light was much greater than it is today. Heat would then have had plenty of time to cross the Universe since the birth of everything. Another possibility is that there was a long 'pre-Big Bang' era. The Universe would then have had ample time to come to an even temperature much as a bath with hot and cold water in it ends up uniformly warm if left alone for long enough.

However, the majority of cosmologists favour a third possibility. Early on, they say, the Universe was an awful lot smaller than we naively infer by simply running the movie of its history backwards in time. Because it was much smaller, heat could easily have crossed from one side of the Universe to the other, ironing out any unevenness in temperature.

The Universe, of course, had to achieve its present size in 13.7 billion years. If it started out smaller, it could only have accomplished this if it expanded faster than expected in the beginning. This is what cosmologists believe. In the first split-second of its existence, cosmologists believe the Universe underwent a brief period of super-fast expansion, dubbed 'inflation'. The precise details of inflation are esoteric and, frankly, not very well understood.* However, most are agreed on what caused the period of super-fast expansion of the Universe: the vacuum.

---

* Inflation may have been driven by a collision between 'island universes' if our Universe is a 'four-brane' adrift in a ten-dimensional space. See Chapter 3, 'Yoyo Universe'.

The Remarkable Properties of the Quantum Vacuum

According to quantum theory, the vacuum is not empty at all. It is seething with restless energy. Energy is permitted to appear out of nothing – in total violation of the principle of 'conservation of energy', one of the cornerstones of physics, which states that energy can be neither created nor destroyed, only changed from one form to another. The proviso is that it pops into existence and disappears again within a very short interval of time. It is a bit like it being OK for a teenager to borrow their dad's car overnight just as long as it is back in the garage early the next morning before he notices it's gone. If energy is borrowed and paid back quickly enough, the law of conservation of energy does not notice.

The continual appearance and disappearance of energy in this way means that the vacuum is in ceaseless turmoil and, on average, contains more than the zero energy naively expected. And the vacuum also exerts a 'pressure', much like the air in a balloon exerts a pressure on the fabric of the balloon. It is this vacuum pressure that is the key to understanding what drove inflation.

According to Einstein's theory of gravity – the general theory of relativity – gravity is generated by two things: energy, of which the energy of mass is the most familiar kind, and pressure.* To be a little more specific, the gravity of a material depends on its energy density – how much energy is crammed into each tiny volume of the material – plus three times the pressure exerted by the material.

The pressure 'term' has been largely ignored since Einstein came up with his theory of gravity in 1915. And with good reason. The pressure exerted by normal matter is completely negligible compared with its

---

* It was Einstein who, in 1905, discovered that mass is just another form of energy – like heat energy or sound energy. Mass energy is the most concentrated of all forms of energy.

energy density. But the possibility has always existed that the Universe might contain hitherto unknown 'stuff' whose pressure is not negligible at all. Enter the vacuum at the beginning of the Universe. According to the proponents of inflation, the vacuum possessed an enormous pressure – in fact, a pressure which was both enormous and negative.

'Negative pressure' sounds a weird concept but, in fact, it is just the opposite of normal, positive, pressure. Stuff with positive pressure wants to expand, like the air in a balloon. Stuff with negative pressure wants to shrink. If it were possible to fill a balloon with it, the fabric of the balloon would simply be sucked inwards rather than blown outwards.

Material with a negative pressure can have a remarkable consequence in Einstein's theory of gravity. Recall that the source of gravity is energy density plus three times the pressure. This means that, if the pressure of a material is negative and big enough, it can completely cancel out the energy density, nullifying gravity altogether. What is more, if the pressure is negative and bigger still, things get even weirder. The 'sign' of the gravity-generating term in Einstein's theory actually reverses. What this means is that, instead of sucking, gravity blows!

Repulsive gravity, it turns out, was the defining characteristic of the 'false vacuum' which existed at the very beginning of the Universe. It was the ultimate driving force behind the super-fast expansion of inflation.* But repulsive gravity was only the beginning. The false vacuum possessed an even more astonishing property.

* It seems impossible that a vacuum, which is everywhere trying to shrink, can actually cause the Universe to expand, or inflate. However, the pressure of the vacuum has no direct effect on the Universe because it is the same everywhere. Every piece of the vacuum is trying to shrink but is surrounded by other pieces of the vacuum that are similarly trying to shrink. Consequently, there is a perfect balance everywhere and the vacuum does nothing but sit still. However, the pressure of the vacuum has an indirect effect on the Universe. Through Einstein's general theory of relativity, it generates the repulsive gravity that speeds up the expansion of the Universe.

11

As the Universe ballooned in size during inflation, the energy density of the vacuum stayed doggedly the same. It meant that, when the Universe doubled in size, the total energy in the vacuum doubled. When the Universe tripled in size, the energy in the vacuum tripled. In fact, for as long as the breakneck expansion of inflation kept going, the total energy in the vacuum just kept going up and up.

Imagine holding a stack of bank notes between your hands, pulling your hands apart and discovering that ever more bank notes fountain out of nothing to fill the gap. That was how vacuum at the beginning of time was. As the Universe grew, energy was literally conjured out of nothing. Inflation, as many physicists have remarked, was the 'ultimate free lunch'.

Eventually, after the merest split-second, inflation ran out of steam. No one knows why or how. But the false vacuum 'decayed'. It transformed itself into normal, well-behaved vacuum.

A split-second, by human standards, is brief. But so violent was inflation that, in that split-second, the Universe grew phenomenally in size, doubling and redoubling its volume perhaps as many as eighty times over. Consequently, by the time inflation came to an end, there was an enormous amount of energy in the false vacuum.

As pointed out above, energy can be neither created nor destroyed, only transformed from one form into another. The energy of the false vacuum therefore had to go somewhere. Where it went was into creating matter and heating it to a ferocious temperature. In short, it generated the blisteringly hot inferno we have come to call the Big Bang.

If the Big Bang is compared with the explosion of a stick of dynamite, the brief epoch of inflation that preceded it can be likened to the explosion of a hydrogen bomb. A billion billion hydrogen bombs. In fact, so violent was inflation that no adequate words exist to describe it.

Inflation and the Never-ending Universe

With inflation bolted onto the basic Big Bang idea, there emerges a new picture of the origin of the Universe. In the beginning, it turns out, was the false vacuum. Driven by repulsive gravity, it underwent a period of extraordinary expansion. The false vacuum was inherently unstable, however, and eventually decayed into normal vacuum. It did not decay everywhere at once, though. The process was far more chaotic than that. Instead, the false vacuum decayed unpredictably at widely separated locations while all the time it was continuing to inflate.

It may help to picture the false vacuum as a vast liquid with tiny bubbles forming spontaneously all over it. The bubbles were regions where inflation had come to an end and the false vacuum had decayed. One such bubble contained our Universe. The energy dumped into this bubble-universe from the false vacuum created matter and heated it to a tremendous temperature. It created the fireball of the Big Bang.

But our Universe was not alone. All the other bubbles continually forming all over the liquid also contained universes. And the energy dumped into these other bubble-universes also created matter and heated it to a tremendous temperature. It drove their very own Big Bangs.

It is a stupendously grand vision of creation. If it is correct, as many cosmologists believe, then the Big Bang was not a one-off event. It was simply one Big Bang among an uncountably huge number of Big Bangs going off like firecrackers across the length and breadth of false vacuum.

People thought the Universe was immense. But, if inflation is correct, it was far more immense than anyone imagined. Douglas Adams was *so* right in *The Hitchhiker's Guide to the Galaxy*, when he wrote: 'Space is big. You just won't believe how vastly, hugely, mind-bogglingly big it is.'

One of the most striking features of the inflationary picture is that the new false vacuum is created by the tremendous expansion of inflation far faster than it can ever decay into normal vacuum. So, despite being constantly eaten away, the false vacuum is never destroyed. Quite the opposite. It keeps on growing and growing. If money behaved like the false vacuum, it would accumulate in your bank account far faster than you could spend it. No conceivable spending spree you could embark on, no matter how extravagant, could prevent you becoming richer and richer. For this reason inflation is unstoppable. It is eternal.

According to 'eternal inflation', our bubble-universe is hopelessly lost in a vast and ever-growing ocean of false vacuum. Although there are other bubble-universes out there in the void, the ever-inflating false vacuum which separates us from them is remorselessly driving them farther and farther away from us. There is no way, not even in principle, that we could communicate with other bubble-universes or even have the slightest knowledge of them.

One of the most remarkable features of bubble-universes like ours is how big they appear to their occupants. Although each bubble-universe has an edge – it is a bubble, after all – to all intents and purposes the space inside extends for ever.

This takes a bit of getting your head around. But it is of key importance for understanding why it is you have a double – in fact, an infinite number of doubles.

The important thing to realise is that, although our bubble-universe is a piece of decayed false vacuum which is no longer inflating, it is nevertheless surrounded by false vacuum which is continuing to inflate at breakneck speed. From where we are sitting, the inflation of that vacuum is happening faster than the speed of light. This means that the boundary between our bubble-universe and the false vacuum must also be receding faster than light.* Since it is impossible for any material object to travel faster than light, the boundary is unreachable – even in principle – by us or any other occupants of our bubble-universe. And, if it is unreachable, the edge of our bubble-universe, for all practical purposes, is an infinite distance away.

This appears to be confirmed by observations of the cosmic background radiation. Slight variations in the temperature of the afterglow of the Big Bang from place to place in the sky turn out to be sensitive to the type of Universe we live in. And the variations appear to strongly favour a Universe that marches on for ever in all directions.

Despite the evidence from the Big Bang radiation, however, our Universe does not 'look' infinite. Far from it. With our most powerful telescopes we can pretty much see all the way to the 'edge' of space and make a rough count of all the galaxies – at last count about 100 billion. But the edge turns out not to be the real edge of our Universe, only the edge of the 'observable' Universe. And this is only a tiny portion of our bubble-universe.

Our view of the bubble-universe is restricted by two things. First, light, though it travels extremely fast, does not travel infinitely fast. And, second, the Universe has not existed for ever but was 'born' a mere 13.7 billion years ago. Taken together, these things mean that the only objects we can see are those whose light has taken less than 13.7 billion years to reach us. Objects farther away than this we cannot yet see because their light is still on its way to Earth. There are whole hosts of objects, for

---

* Though matter and energy cannot travel faster than light, according to Einstein's general theory of relativity, space – the backdrop against which the cosmic drama is played out – can expand at any rate it likes. Inflation is a prime example of this faster-than-light expansion. After all, it ensured that the Universe at 450,000 years old was far bigger than 450,000 light years across, the maximum size it could have attained if its expansion had been limited merely to the speed of light.

instance, whose light would take 14.7 billion years to travel across space to us. However, they will not become visible from Earth for another billion years.

Actually, this is not completely true. Because the expansion of the Universe in the past was faster than it is today, the 'light horizon', which defines the edge of the observable Universe, is farther away than 13.7 billion light years. It means that we can see the light from all objects that are closer than 42 billion light years away (4×10^{26} metres).

The Universe's horizon has much in common with the horizon seen from a ship at sea. Just as we know there is more sea over the horizon, we know there is more of the Universe over the cosmic horizon – in fact, an infinite amount, according to inflation.

Of course, our observable Universe is not the only region in our bubble-universe bounded by a light horizon. If the space within our bubble-universe is effectively infinite in extent, it follows that there must be an infinite number of similar regions, each bounded by their own horizon. So, what is it like in these other regions? Here, quantum theory has something very startling to say.

Quantum Theory and the Graininess of Universe

According to quantum theory, matter is not continuous but grainy. If you picked up a branch or a rock and cut it in half, then in half again, you could not go on in this manner for ever. Sooner or later, you would come to a tiny, hard grain of matter that could not be divided any more. In the 5th century BC, the philosopher Democritus called such a grain an 'atom', from the Greek words *a-tomos* meaning 'uncuttable'. Nowadays, we know that the truly uncuttable motes of matter – quarks and electrons – are even smaller than atoms. But this does not change the central truth. Matter, on the smallest scales, is fundamentally grainy, like a newspaper photograph looked at too closely.

And it is not just matter. According to quantum theory, everything – energy, space, even time – ultimately comes in tiny, indivisible chunks, or 'quanta'.

The basic building blocks of today's Universe – the fundamental grains of matter – are protons and neutrons.* Essentially identical in size, these two particles are the principal constituents of the 'nuclei' at the heart of atoms.

The protons and neutrons of the Universe are distributed very unevenly. For instance, there is a whole bunch of them over here making up the Earth and a whole bunch over there constituting the Sun. What is more, the protons and neutrons which make up the Earth and Sun are spread very thinly. This is because atoms – apart from the tiny, hard motes of their nuclei – are overwhelmingly made of empty space.

No one has provided a better mental picture of an atom than the playwright Tom Stoppard in *Hapgood*: 'Now make a fist, and if your fist is as big as the nucleus of an atom then the atom is as big as St Paul's, and if it happens to be a hydrogen atom then it has a single electron flitting about like a moth in an empty cathedral, now by the dome, now by the altar.'

It is easy to imagine other possible universes in which the protons (and neutrons) of ordinary matter are distributed in different ways. A pointless exercise, you might think. And you would be right if the other possible universes were merely hypothetical. But they are not!

---

* Protons and neutrons are actually triplets of quarks bound together. This is because quarks, despite being the ultimate indivisible grains of matter, are not able to roam free in today's Universe. This was possible only in the super-hot conditions which existed in the first split-second of the Big Bang.

Consider the question of how the galaxies – the dominant structures in today's Universe – came to be. Inflation, it turns out, provides an answer.

In the beginning, recall, was the quantum vacuum. Now, it is a fundamental property of anything quantum that it is seething with restless energy. Consequently, on the smallest scales, the quantum vacuum was like a boiling cauldron. (On large scales everything averaged out. The quantum vacuum was like an ocean that looks flat from an airliner flying high above, but close up is seen to be choppy.)

The choppiness of the quantum vacuum should have been of no consequence. It was a feature, after all, of impossibly tiny regions. But inflation changed everything. Not only did it inflate the vacuum, it inflated the choppiness of the vacuum too. As a result, the choppy regions became stretched to tremendous size.

Now, the choppiest regions of the vacuum contained the most energy. And energy, as Einstein recognised in his theory of general relativity, warps space, creating gravity. Warped space and gravity are one and the same thing. So, when inflation finally spluttered to a halt, and the energy of the false vacuum was dumped unceremoniously into matter and normal space, it was the regions where the vacuum was choppiest that became the places where gravity was strongest. And, on account of their enhanced gravity, these regions were the most effective at dragging in and piling up matter from their surroundings. They would grow into the great clusters of galaxies which we see snaking like daisy chains across today's Universe.

The inflationary picture therefore tells us that giant collections of trillions upon trillions of suns were actually 'seeded' by regions of the vacuum far smaller than an atom!

But the quantum contortions of the primordial vacuum that spawned galaxies were more than just small. They were utterly random. This may not seem very significant. But it is. It means that everything in the observable Universe – the distribution of galaxies seen by our telescopes – is ultimately the result of random processes that went on in the first split-second of the Universe's existence. And what is true of our observable Universe must also be true of all other regions the size of the observable Universe in our bubble-universe. The way their galaxies are distributed must also be the result of random processes in the first split-second of the Universe.

This is crucial. It means that all the possible arrangements of protons in a volume the size of our observable Universe are not merely hypothetical. They will in fact occur – in other regions of the bubble-universe.

There is of course a stupendously large number of ways that protons could be arranged in a volume the size of the observable Universe. But the hugeness of this number is not important. The important thing is that it is finite. This follows from the fact that protons are tiny grains so there is only a finite number of places in space where they can be located, and a finite number of ways they can be arranged. Think of a chessboard with a limited number of squares where chess pieces can be located and a limited number of ways of arranging those pieces. According to quantum theory, the observable Universe is like a giant three-dimensional chess board.

Because there is only a finite number of ways of arranging protons in a volume the size of our observable Universe, and the bubble-universe, according to inflation, is effectively infinite in extent, it follows that every possible arrangement must occur somewhere. In fact, it is even more incredible than this. Every possible arrangement must occur an infinite number of times, in an infinite number of other places.

The implication for our own observable Universe is as obvious as it is shocking. It cannot be unique. Worst still, there must be an infinite number of other regions exactly like our observable Universe.

How Far Away are Regions Identical to Ours?

How far away is the nearest region that is identical in all respects to our observable Universe? Well, this depends on how many different Universes are possible. Which is, of course, the same as asking how many different ways protons can be arranged within a volume the size of our observable Universe.[*]

Think of protons as being like tiny oranges that can be stacked together, row on row, layer on layer. Just as it is possible to calculate how many oranges will fit in a box, it is possible to estimate how many protons will fit in the 'box' of the observable Universe. The answer turns out to be about 10^{118}.[†]

The observable Universe is not, of course, chock-a-block full with protons and neutrons. Nevertheless, because of its quantum graininess, it has 10^{118} distinct locations, each of which may or may not be occupied by a proton.

In estimating how many different ways in which protons can be distributed around these 10^{118} locations, it helps first to consider a drastically cut-down Universe with a more manageable number of locations for protons – say four. Location 1 can either contain a proton or no proton, making two possibilities. For each of these possibilities, location 2 can either contain a proton or no proton. That makes a total of $2 \times 2 = 4$ possibilities. For each of these, location 3 can have either a

[*] For this elegant argument I am indebted to Max Tegmark of the Massachusetts Institute of Technology ('Parallel Universes', *Scientific American*, May 2003, p 31).
[†] There are actually only about 10^{80} protons in the observable Universe, which gives an indication of how mind-bogglingly thinly matter is smeared through space. Each atom is so empty that there actually is room for another million billion protons inside it. And so empty is the observable Universe that there is room for another ten thousand billion atoms. In fact, there is so much empty space about that it is as if we live in a ghost Universe!

proton or no proton. That makes $2 \times 2 \times 2 = 8$ possibilities. By now, the pattern should be clear. For a universe with four possible locations, protons can be arranged in $2 \times 2 \times 2 \times 2 = 2^4$, or 16, different ways.

It is easy to generalise. If there are n possible locations for protons, the protons can be arranged in 2^n different ways. In other words, a total of 2^n different universes are possible. In our Universe, $n = 10^{118}$, so there are $2^{10^{118}}$ possible ways the basic building blocks could have been arranged. This is approximately equal to $10^{10^{118}}$.[*]

Now we are in a position to address the original question. How far away is the nearest region which is identical in all respects to our observable Universe?

Think again of that cut-down universe with only four locations for protons. There are $2^4 = 16$ possibilities, after which universes start repeating. If these are arranged in a three-dimensional space, the first repeating Universe is only twice the width of the Universe away. In general though, the distance is $2^{(n-3)} \times$ the diameter of the observable Universe. Since the diameter of the Universe is about 8×10^{26} metres, the nearest region identical to our observable Universe is about $10^{10^{118}}$ metres away.[†]

This is a fantastically long way away. However, things are not quite this bad. Although we would have to travel $10^{10^{118}}$ metres to find a region identical to our observable Universe, it would not be necessary to travel so far to find a region identical to our local neighbourhood. Take,

---

[*] $2^{10^{118}} \sim (10^3)^{10^{117}} = 10^{(3 \times 10^{117})} \sim 10^{10^{118}}$ (where '$\sim$' means approximately equal to)

[†] For $n = 10^{118}$ and $d = 8 \times 10^{26}$ metres,

$$D = 2^{(10^{118}-3)} \times 8 \times 10^{26} \text{ metres}$$
$$\sim 2^{10^{118}} \times 8 \times 10^{26} \text{ metres}$$
$$\sim 10^{10^{118}} \times 8 \times 10^{26} \text{ metres (see previous footnote)}$$
$$= 10^{10^{118}} \times 10^{26} \text{ metres}$$
$$= 10^{[10^{118} \times 26]} \text{ metres}$$
$$\sim 10^{10^{118}} \text{ metres}$$

for instance, a sphere 100 light years across and centred on the Earth. How far would we have to travel to find another identical sphere?

Well, a sphere 100 light years across is roughly a billionth the diameter of the observable Universe, which means it contains only a billion billion billionth the number of possible locations for protons, which comes to 10^{91}. That means that, instead of there being $10^{10^{118}}$ possible arrangements, there are only about $10^{10^{91}}$. So, after $10^{10^{91}}$ spheres 100 light years across have been exhausted, things start to repeat. That puts the nearest 100-light-year sphere of space identical to our own at about $10^{10^{91}}$ metres from the Earth.

That is still a fantastically long way away. But your nearest double is still not that distant. All that is really necessary is an assemblage of matter identical to you. Roughly speaking, that means about 10^{28} particles. So, how far would we have to travel to find another identical assemblage? Well, if there are 10^{28} possible locations for protons, there are $10^{10^{28}}$ possible arrangements of those protons. And this means your nearest double is just $10^{10^{28}}$ metres away.

In reality, your nearest copy is probably much closer. This is because the processes of planet formation and biological evolution may conspire to create planets like Earth and intelligent creatures like humans. Astronomers suspect that there may be at least 10^{20} habitable planets in the volume of the observable Universe. And some of them may look like Earth.

Could You Ever Meet Your Double?

Could you ever meet your double? Remarkably, there is no problem – at least in principle. After all, the horizon of the Universe moves outwards by roughly a light year every year, so the observable Universe gets ever bigger, eating ever more into the bubble-universe. It is easy to imagine a time when the volume of our observable Universe finally overlaps an identical region.

Unfortunately, because of the enormous distances currently separating identical regions, our observable Universe will not merge with its double until far, far into the future. By that time, the human race will be long gone and all the stars will have gone out. So, for all practical purposes, it will not matter one way or another whether it happens or not.

There is another possibility, however. Similar regions may never overlap. Currently, the expansion of the Universe is speeding up because of the presence of dark energy. If dark energy continues to speed up the expansion, then it will become more and more difficult for light from distant galaxies to reach us. Even as the light is en route, the space between us and the galaxies will be stretching. A light beam trying to reach the Earth will be in the position of a 100-metre runner trying to reach the finishing line when the finishing line is continually being moved farther away.

The effect will be to cause the horizon of the Universe eventually to stop moving outwards into the bubble-universe and begin moving inwards. The observable Universe, instead of growing, will begin shrinking. If this is the case, our observable Universe will never overlap with a similar region – not even in principle.

However, we do not know for sure that this is what will happen. This is because the expansion of the Universe began speeding up only relatively recently. Nobody has the slightest idea why. And, since they do not know why, it follows that nobody has any idea whether the expansion will continue speeding up for ever or whether it will one day run out of steam.

All Possible Histories and the Many Worlds

Since each possible universe evolved to its present state from a different choppy state of the quantum vacuum, saying that all possible universes exist somewhere in the bubble-universe is entirely equivalent to saying that all possible histories, from the Big Bang to the present day, are played out somewhere.

In the huge majority of universes, neither the Sun nor the Earth ever arose. However, there must be other universes containing Earths. And, among those, there must be Earths which were not devastated by a comet impact 65 million years ago and on which the dinosaurs evolved into intelligent beings. There must be universes with Earths where Rome was defeated by Carthage, where Einstein became a violin teacher and not a physicist, and where the Nazis prevailed in the Second World War. There must be a universe where Elvis is alive and well. In fact, Elvis must be alive in an infinite number of other space domains!

Those who are familiar with quantum theory may have noticed a certain similarity here with the so-called Many Worlds interpretation. Like all such 'interpretations', this is an attempt to explain why quantum theory predicts atoms can be in many places at once whereas such a thing is never observed. According to the Many Worlds, however, atoms can be in different places at once – but in different realities, or parallel universes. The Universe keeps bifurcating. If a quantum event can happen one way or another way – say, an atom can disintegrate or not – the Universe splits into copies, one in which the first outcome is played out and another in which the second outcome occurs. Because so much of what happens in the world hinges on quantum events – for instance, the random event that causes cancer through a mutation in DNA – all possible histories of the Universe are played out.

The Many Worlds seems to be very wasteful of universes. However, it faithfully predicts the outcome of all known experiments. Nevertheless, the location of the other universes of the Many Worlds interpretation is never specified. Nobody has the slightest clue where they might be.

In marked contrast, the other universes of the standard theory of cosmology are utterly concrete. We know exactly where they are. So many metres in that direction. If you believe the standard picture, you have to believe in them.

Their existence provides a major philosophical and moral headache. After all, if all possibilities happen somewhere, what's the point of fighting for a better world? No matter how successful you are, there will always be other Earths – in fact, an overwhelming number of them – where copies of you fail.

The science fiction writer Larry Niven explored this idea in his story 'All the Myriad Ways'.[*] In Niven's world, the Crosstime Corporation has made billions by importing and patenting scores of inventions from alternative time tracks. But the company's founder has jumped from the balcony of a thirty-sixth-floor luxury apartment, the latest in a series of inexplicable suicides which began only a month after Crosstime started. A detective embarks on an investigation and gradually realises the truth. People have committed suicide because of the knowledge of the other versions of themselves, the might-have-beens that lived lives that were less lonely or more fulfilled. They have committed suicide because of the knowledge that, if all possibilities happen, nothing you do ever matters; whatever decision you make, the opposite decision will also be made in some other reality. They have committed suicide out of despair.

If the existence of all the other yous seems extremely difficult to take, remember that it is an unavoidable consequence of the standard theory of our Universe, which is embraced by just about anybody who is

[*] See my book *The Universe Next Door* (Headline, 2002).

anybody in science. The standard picture incorporates both quantum theory and inflation. So, if it is not true, then either quantum theory or inflation is incorrect. It would be nigh on impossible to find a physicist who believes quantum theory is fundamentally wrong or a cosmologist willing to discard inflation. But there it is. Either you throw away quantum theory or bin inflation. Or you are not the only version of you reading this – and Elvis is still alive and well in another space domain.

2

Cosmic Computer

Where does the complexity of the Universe come from? A simple computer program is generating it!

> There is something fascinating about science. One gets such wholesale
> returns of conjecture out of such a trifling investment of fact.
> Mark Twain, *Life on the Mississippi*, 1884

> God has chosen the world that is the most simple in hypotheses
> and the most rich in phenomena.
> Gottfried Leibniz, *Discourse de métaphysique*, 1686

It's AD 2068 and the survey expedition from Earth is picking its way through the ruins of an alien civilisation, long departed from its home world for who knows where. Ahead, bathed in the sombre light of the twin red suns, is a great slab of a building – the planet's central library, repository of the civilisation's accumulated wisdom.

Struggling visibly in the strong gravity, the expedition members clamber up the giant steps and push open the creaking door. Their boots reverberating in the thick atmosphere, they hurry through an empty, echoing chamber – until, finally, they come to a single cabinet, displaying a lone tablet inscribed with arcane symbols. Everyone crowds around while someone scans it with a translator ...

RECIPE FOR UNIVERSE:

27

RUN COMPUTER PROGRAM (BELOW)

REPEAT FOR 13.7 BILLION YEARS

One person laughs. Another gasps in disbelief. The cosmic computer program they are all staring at is only four lines long.

Could the recipe for making a universe really be as simple as this? One present-day physicist is convinced of it. His name is Stephen Wolfram and he claims to have stumbled on nature's 'big' secret. The source of all its bewildering complexity – from spiral galaxies to rhododendrons to human beings – is the application of a few simple instructions, over and over again. 'Our Universe is being generated by a simple computer program,' says Wolfram.

Wolfram, a child prodigy from London, began publishing papers in professional physics journals at the age of fifteen. What led him to his extraordinary conclusion is a discovery he made around 1980. Contrary to all expectations, he found that simple computer programs have the ability to generate extraordinarily complex outputs.

Wolfram's discovery came about when he became interested in problems such as how galaxies like our Milky Way form and how our brains work. 'The trouble was that none of these "complex systems" seemed explicable by conventional science,' he says.

Conventional science is synonymous with maths-based science. In the seventeenth century, Isaac Newton discovered that the laws which govern the motion of a cannon ball through the air and a planet round the Sun could be described by mathematical formulae, or 'equations'. Following Newton's lead, generations of physicists have found that mathematical equations exist that can perfectly describe everything from the character of the light given out by a hot furnace to the warping of space and time by the concentrated mass of a black hole.[*]

* A black hole is a region of space where gravity is so strong that not even light, the fastest thing in the Universe, can escape.

But, despite the tremendous successes of equation-based science in penetrating nature's secrets, it has an Achilles' heel: it cannot do 'complexity'. It is utterly incapable of capturing the essence of what is going on in a whole range of complex phenomena, ranging from turbulence in fluids to biology itself.

Most scientists lose little sleep over this. Complex phenomena may be 'hard', they say, but this does not mean that science will not eventually get round to tackling them. Wolfram, however, emphatically disagrees. Controversially, he believes that mathematical science will never, ever penetrate the mystery of complex phenomena.

A streetlight illuminates merely what it can illuminate – the circle of ground immediately beneath. Similarly, Wolfram believes mathematical science illuminates merely what it is capable of illuminating – those phenomena whose essence can be captured by mathematical equations. But such phenomena, he contends, are rare and unusual. In the same way that a streetlight fails to reveal the subways and sports grounds and art galleries of its surrounding city, science as practised for the past three centuries is blind to the overwhelming majority of phenomena in the Universe – complex phenomena. 'Since such phenomena include living things, the human brain and the biosphere, we are talking about all the truly interesting things that are going on in the Universe,' says Wolfram.

This is radical stuff. For centuries, physicists have wondered why nature obeys mathematical laws, which can be distilled into neat mathematical equations and which can then be scrawled across black-boards. The Hungarian-American physicist Eugene Wigner famously drew attention to this when he talked of 'the unreasonable effectiveness of mathematics in the physical sciences'.

According to Wolfram, however, Wigner was wrong to believe that the Universe is essentially mathematical. Mathematics is no more effective in revealing the inner workings of nature than a streetlight is in revealing the city that surrounds it. Nature may 'appear' to follow mathematical laws, he says. However, that is hardly surprising when scientists specifically seek out the rare natural phenomena that follow mathematical laws. 'Wigner could equally well have remarked on the unreasonable effectiveness of streetlights in illuminating the ground beneath them,' says Wolfram.

If Wolfram is right, science has a serious problem. After all, if mathematical equations are incapable of describing nature's most interesting phenomena – complex phenomena – how can such phenomena be described? In the early 1980s, Wolfram gave this question a great deal of thought. It was clear to him that the Universe must obey rules of some kind. If it did not, after all, there would be no pattern or regularity in nature. The Universe would be a meaningless maelstrom of unpredictable randomness and chaos. But, if the rules are not embodied in mathematical equations, what are they embodied in? It was clear to Wolfram that it had to be something more general than a mathematical equation. After thinking about it, he could come up with only one thing that fitted the bill: a computer program.

Nature's Big Secret

Wolfram decided to find out what kind of science could be built starting with the more general kinds of rules embodied in computer programs. The first big question he needed to answer was: what are such rules capable of? Or, to put it in another way, typically what do simple programs do?

The simplest computer program Wolfram could think of is known as a 'cellular automaton'. The most basic of these is simply a long line of squares, or 'cells', drawn across a page. A cell can be one of two colours – white or black. At regular intervals of time, a new line of cells is drawn on the page, immediately above the first. Whether a cell in this second line is black or white depends on a rule applied to its two nearest neighbours in the first line. The rule might, for instance, say: 'If a particular cell in the first line has a black square on either side of it, it should turn black in the second line'. A third line of cells, immediately above the second, is then created by applying the cellular automata 'rule' to the second line, and so on.

What we are talking about here is the operation of a simple computer program embodying the cellular automaton rule. The program takes an input – the pattern of black cells and white cells on one line – and produces an output – the pattern of cells on the next line. The key thing is that the output is fed back in as the next input to the computer program, rather like a snake swallowing its own tail. Such tail-swallowing is commonly called 'recursion'. And, as a wit once said: 'To understand recursion, you must first understand recursion!'

For a one-dimensional, two-colour, adjacent-cell cellular automaton like this, it turns out there are 256 possible rules, 256 kinds of program.*
The question is: what happens when the programs are run, starting, say,

---

* Why 256 possible rules? Well, for each successive line, the colour of a cell (black or white) depends only on its own previous colour and the colour of the cell on the left and the cell on the right. This means there are eight possible starting situations. For instance, a square can be black with a black square to its left, or black with a black square to its right, or black with black squares to the left and right, or black with white squares to the left and right. Another four possibilities arise if the central square is white.

Each rule maps all these eight input situations to an output (black or white). This means there are $2^8 = 256$ possible rules for such a one-dimensional, two-colour, adjacent-cell cellular automaton.

with a single black cell in the first line of cells? In true scientific fashion, Wolfram began experimenting to find out.

He soon discovered that some rules and some starting patterns led to nothing interesting. As new lines of cells were created, any pattern quickly fizzled out. Or a particular arrangement of black and white cells began repeating endlessly. However, in some cases, something very much more interesting happened.

The early 1980s was the time of the first cheap desktop computers so Wolfram was able to watch his cellular automata perform on a computer rather than on a piece of paper. Seeing the new lines of cells marching steadily up the screen was much like watching a movie. Occasionally, the patterns of black cells coalesced into discrete 'objects'. These persisted – as unchanging and stable as a table or chair – despite the fact they were being continually destroyed and regenerated.

Wolfram played with his cellular automata for hours on end, mesmerised by the marching patterns. And then, one day, he stumbled on something extraordinary. 'I found a pattern which appeared never to repeat, no matter how long I stared at it,' says Wolfram.

If you see a complex thing like a car or a computer, you know it must have been made by a complex process. Even in biology, where natural selection is blind, the complexity of organisms is a result of a complex series of processes operating over billions of years of evolution. In the everyday world, simple things have simple causes and complex things have complex causes. What Wolfram had found, however, was something that bucked the trend – a complex thing that had a simple cause.

For Wolfram it was a life-changing moment. As he stared at his computer screen and the never-ending novelty scrolling down it, he wondered: Is this the origin of the Universe's complexity? 'When nature creates a rose or a galaxy or a human brain, is it merely applying simple rules – over and over again?' he asks. 'Is this its big secret?'

A Survey of All Possible Worlds

From that moment on Wolfram became obsessed with the origin of complexity. At the time of his epiphany he was at the California Institute of Technology in Pasadena. However, in the mid-1980s, he moved first to Princeton's Institute for Advanced Study – Einstein's old institute – then to the University of Illinois at Urbana-Champaign, where he founded the Center for Complex Systems Research. Around the same time, he started the first scientific journal on complexity. He even created his own computer language – 'Mathematica' – which helped him in his investigation of the origin of complexity.

Mathematica led him to start his own company, Wolfram Research, and attract scientists and mathematicians to help develop the software. The programming language turned out to be not only a tool but an inspiration to his work. Although Wolfram assembled it from simple program 'modules', it was nevertheless capable of carrying out enormously complex tasks. 'It hammered home to me once again my central discovery – that simple programs can have hugely complex outcomes,' says Wolfram.

Wolfram's hope was that others would pile into the research area he had created and that this would lead to rapid progress in understanding complexity. To his disappointment and frustration, however, few joined in and progress was slow. He became increasingly impatient. By early 1991, he decided there was only one thing to do – carry out the work himself.

With several million users worldwide, Mathematica had made Wolfram a multimillionaire. He did not need to be employed by a university and he did not need to fight constantly for research money. He was free to concentrate all of his time on creating a science of complexity.

Wolfram had set himself a gargantuan task but even he did not realise it would take him a decade. During that time, he published not a single research paper. Although he certainly talked and corresponded with other scientists, he pretty much vanished off the edge of the scientific radar screen.

Month after month, year after year, while the rest of the world slept, Wolfram laboured through the night, painstakingly laying the foundations of a new way of doing science. In essence, he was carrying out a systematic computer search for simple rules with very complicated consequences. 'He set out to survey all possible worlds — at least all the ones generated by simple rules,' says the mathematician Gregory Chaitin of IBM in Yorktown Heights, New York. 'The result was a treasure trove of small computer programs that, when repeated again and again, yield extremely rich, complicated and interesting behaviour.'*

Among the many things Wolfram discovered is the remarkable property of cellular automaton rule 110. Starting with a single black cell, this simple rule turns out to be capable of generating infinite complexity, infinite novelty, infinite surprise. Not only that but a cellular automaton following rule 110 is a 'universal Turing machine'.† Despite being amazingly simple, it is like a modern-day computer that can carry out any imaginable computation, simulate any other conceivable machine.

The remarkable ability of cellular automaton rule 110 is highly suggestive. After all, if even a simple one-dimensional cellular automaton can create never-ending complexity, it shows the kind of power that nature potentially has at its disposal. And Wolfram is convinced that nature avails itself of that potential. 'I believe that physical systems subject

---

* For more on Chaitin — in fact, for a whole chapter on the man, not to mention the amazing number he invented that contains the secret of life, the Universe and everything, see Chapter 6, 'God's Number'.
† See Chapter 6, 'God's Number'.

to simple rules applied recursively – with the output fed back in as the input – can have created everything from the tip of your nose to the most distant cluster of galaxies,' he says.

So is the Universe a giant cellular automaton – a three-dimensional version of the one-dimensional ones Wolfram has been playing with on his computer? Surprisingly, Wolfram thinks not. 'I think the truth is actually much more strange and interesting,' he says.

The Universe-Generating Program

A serious shortcoming of a cellular automaton as a model of the Universe is that all the cells update themselves together. This kind of coordinated behaviour requires a built-in 'clock', whose ticks provide the all-important cue for the cells to 'all change'. Unfortunately, this kind of clock is impossible to implement in the real Universe, the reason being the existence of a cosmic speed limit, as discovered by Einstein.

Nothing, it turns out, can travel faster than the speed of light. This constraint means that, wherever the cellular automaton clock happens to be located in the Universe, the signal carrying news of its tick will take longer to travel to a cell that is far away from it than to one that is nearby. A possible way round this might be to have lots of clocks distributed throughout the Universe. However, this does not overcome the fundamental problem because there is no way to make sure the clocks are all telling the same time. If a 'reference clock' is used, inevitably the signal carrying news of its time will take longer to reach some clocks than it does to reach others.

The impossibility of implementing a 'global' clock in our Universe means that at the very least the Universe cannot be a 'standard' cellular automaton. However, that does not rule out its being a cellular automaton of some non-standard type – one that somehow gets by without a global clock. This seems a bit of a tall order. But Wolfram can think of an ingenious way it can be achieved. 'Say that, rather than updating all of its cells together, a cellular automaton updates just one cell at a time,' he says.

At first sight this may seem crazy. But consider for a moment the advantage of such a scheme. If, at each step, only one cell is updated, the sticky problem of getting all the cells to update at the same time clearly goes away.

Of course there remains the small matter of how such a cellular automaton could possibly mimic our reality. After all, we have the very strong impression that everything in the Universe is travelling forward through time together, not that one detail of reality is being updated in turn while everything else remains doggedly rooted to the spot.

Say you are playing in a football game. You see all the other players running about the pitch simultaneously. You do not see first one player take a step, as their thought processes click on one notch, while everyone else remains paralysed in mid-stride; then another player take a step, and so on. 'But, just because you do not see this happening, does not mean that this is not exactly what is going on,' says Wolfram.

But surely you would notice? No, says Wolfram. The only time you notice the world about you is when it is your turn to be updated. And, when this happens, all you see is that all the other players have moved on a fraction. Because your awareness is frozen between your own updatings, it is impossible for you to notice when any of the other players are updated. Despite the fact that only one player on the pitch is moving at any one time, your perception is of everyone on the pitch running about simultaneously.

Between any two successive moments of time as perceived by you, there are very many updating events, none of which you have any awareness of. In fact, all you can ever really know about, says Wolfram, is what updating event influences what other updating event. For instance, the updating event that moved the football one step closer to the opposing team's goal influenced the opposing team's defenders and goal keeper, who altered their positions to intercept the ball. This, of course, is the familiar story of a football game. 'But that's all it is – a *story*,' says Wolfram. 'A network of cause and effect we impose on the underlying reality to make some kind of sense of it.'

Contrary to common-sense expectations, then, it appears that it is possible to mimic our Universe with a cellular automaton in which only one cell at a time is updated. The passage of the 'time' in the Universe is marked by the regular ticking of the cellular automaton's clock. That only leaves 'space' to worry about. Unfortunately, it is here, according to Wolfram, that the idea of the Universe-as-a-cellular-automaton comes to grief.

Wolfram is convinced that the computer program generating our Universe is a simple one. Every scrap of evidence he has accumulated since his key discovery that simple programs can produce unexpectedly complex outputs bolsters this belief. But, if the program generating the Universe is simple, it stands to reason there will not be room in it for much 'stuff'. In other words, very few of the features of our Universe – from gravity to space and time to koala bears – will be visible in the program. Instead, they will 'emerge' – like an inflatable raft unfolding from a canister – only after the program has been running for a long while.

But Wolfram does not simply think the Universe-creating program is simple. He goes further than this. He believes the program may be among the simplest possible programs capable of generating the Universe. This is a leap of faith. All Wolfram knows for sure is that the rule for the Universe is not really complicated. If it was, he argues, there would be no perceptible pattern to nature, which there clearly is. Wolfram thinks it is possible our Universe is the very simplest universe that is not obviously a silly one – for instance, a universe with no notion of space or of time. Consequently, he thinks it is worth first trying the simple rules for size because our Universe might be among them.

If the Universe-generating program is indeed among the simplest programs capable of generating the Universe, it will contain the absolute bare minimum of stuff. And it is this that persuades Wolfram that the Universe cannot possibly be a cellular automaton. A cellular automaton, after all, is a rigid array of cells laid out in 'space'. In other words, the very notion of 'space' is built into its very foundations. To Wolfram, this is already too much stuff.

Wolfram believes the Universe-generating program will be so simple, so pared down, that even something as apparently fundamental as space will not be built into it. Instead, it will emerge along with everything else only as the program runs, conjured out of something even more basic than space.

Wolfram believes space is not a smooth, featureless backcloth to the drama of the Universe. Instead, it has an underlying structure. The analogy he uses is water. Although water looks smooth and continuous, in fact it is made up of tiny motes of matter called molecules. Wolfram thinks space is similar. If it were possible to examine it with some kind of super-microscope, we would see that it is made of a huge number of discrete points. The points, or 'nodes', are connected together in a vast extended network.

But how can a mere network of points have the properties of familiar space? 'Surprisingly easily,' says Wolfram. 'It simply depends on the way the nodes are connected to each other.'

Imagine being at one particular node, then going to all the nodes that are one connection away, then two connections, then three, and so on. After going, say, r connections, simply count how many nodes you have visited. If there are roughly pi $\times r^2$ nodes – the area of a circle – then the space is two-dimensional like the surface of a piece of paper. If there are roughly ⅔ pi $\times r^3$ – the volume of a sphere – then the space is three-dimensional, like the space we live in. It turns out that a simple network of nodes can mimic the essential properties of absolutely any space imaginable, be it one-dimensional, two-dimensional or 279-dimensional.

According to Wolfram, space is nothing more than a bunch of nodes connected together. Of course, there is a little bit more to it than that.

Wolfram envisages a space network being updated in a similar way to a cellular automaton. After all, a constantly updated cellular automaton has a proven ability to generate complexity reminiscent of our Universe. Recall how it was possible to get over the synchronisation problem of a cellular automaton by updating just one cell at a time. Well, Wolfram thinks that this elegant solution can be carried right over to a space network. Instead of having a rule which says, if a cell is surrounded by a certain pattern of coloured cells – change its colour, Wolfram imagines a rule saying, if there is a piece of network with a particular form, replace it with a piece of network with another form. Remarkably, Wolfram claims that everything in our world can emerge from such a space network.

Take particles of matter. In a cellular automaton – for instance, the one subject to rule 110 – the system may quickly organise itself into a few localised structures which are persistent and appear to move through space just like fundamental particles – quarks and electrons and so on. What is actually happening is that, as fast as the structures are destroyed, they are refreshed again. It is just like a TV image of a football game. We may perceive that a football is in flight. But, in reality, what is happening is that a picture of the ball is being refreshed thirty times a second and giving us the illusion of the ball moving through the air.*

Sometimes, in a cellular automaton subject to rule 110, there is a collision between 'particles'. They slam into each other and a whole bunch of other particles come out. This is just the kind of thing physicists observe at atom smashers like the one at the European centre for particle physics at CERN in Geneva. And what happens in a cellular automaton subject to rule 110 can also happen in a space network. Instead of being stubbornly persistent patterns of cells, however, the 'particles' are stubbornly persistent tangles of connections.

Remarkably, Wolfram has found that, with a constantly updated network of nodes, it is possible to create both the space we live in and the matter we are made of. 'Reality,' as Einstein remarked, 'is merely an illusion, albeit a very persistent one.'

A problem arises, however, if a rule applies to a particular pattern of nodes and there are several places in the network with the same pattern. Which place should be updated first? Updating the places in a different order will in general lead to different networks of cause and effect. Rather than having a unique history, the Universe will have several possible histories. We will not know why we are following the history we are and not another, which is a highly unsatisfactory state of affairs.

Fortunately, there is a way out of this difficulty, says Wolfram. By a stroke of luck it turns out that there are certain rules with the property that it in fact does not matter in which order they are applied. Wolfram calls them 'causally invariant' rules. 'Whenever they are used, there is always just a single thread of time in the Universe,' he says.

---

* Even the atoms and molecules that compose you are 'refreshed' at intervals. They are not the same ones that were a part of you last year. Most cells such as blood cells are replenished within a matter of weeks and even those that persist longer such as neurones have their component molecules changed at regular intervals.

Wolfram's progression from the Universe-as-a-cellular-automaton to the Universe-as-a-constantly-updated-space-network is a good illustration of the way in which physicists grope their way towards a true picture of nature. They start with a crude model which mimics an aspect of reality which they consider to be important. In this case, the model is a cellular automaton which can generate complexity tantalisingly like the complexity we see in the world around us. Inevitably, the model falls short in some way. In the case of cellular automata, it contains too much ready-made stuff such as 'space'. Nevertheless, physicists use the crude model as a bridge to reach a better model that mimics more reality more faithfully. Lastly, they throw away the bridge.

Wolfram's talk of space and matter 'emerging' from a network may seem rather woolly. However, he maintains that it can explain concrete things too, such as the general theory of relativity, Einstein's theory of gravity. In a nutshell, the theory says that matter distorts, or warps, space-time, and that warped space-time is what matter reacts to when it moves. In fact, warped space-time is all that gravity is. We think that the Earth pulls on the Moon with invisible fingers of force which somehow reach out across 400,000 kilometres of empty space. But, according to Einstein, this is an illusion. In reality, the Earth's mass warps space-time, creating a sort of valley in its vicinity. We cannot see it because space-time is four-dimensional and we can experience only three dimensions. But the Moon 'sees' it. It skitters around the rim of the valley in space-time like a roulette ball round a roulette wheel.

Wolfram claims that his perpetually updated space network behaves exactly like Einstein's warped space-time. For simplicity, imagine things in two dimensions. Also, imagine that the network is a network of hexagons which can be laid out flat like a fishing net spread out on a beach. What happens if some of the connections are changed so that some heptagons and pentagons are mixed in with the hexagons? The answer is that the network bulges out or in. 'This is warped space,' says Wolfram.

In ordinary, flat, two-dimensional space, as mentioned above, the number of nodes we get by going out r steps through the network goes up as r^2. Well, in a warped network, it is not quite the same. There is what mathematicians call a 'correction term'. And it turns out that the correction term is basically the 'Ricci tensor'. It is not necessary to know exactly what the Ricci tensor is, but it crops up in Einstein's equations, which in general relativity specify the warpage of space-time.

The story of how is quite complicated. But Wolfram maintains that, with just a few assumptions, he can work out the conditions which the Ricci tensor must obey. 'And, guess what?' he says. 'They seem to be exactly Einstein's equations of gravity.'

The Ubiquity of Biological Complexity

Wolfram believes the computer program that nature is using to generate the Universe is very short. We are certainly not talking about the ten million or so lines of a program like Microsoft Windows. Far from it. 'Nature's program may be expressible in as few as four lines of Mathematica,' he says.

If he is right, those lines are responsible for creating everything from chocolate doughnuts to TV game shows to the very thought processes that have led Wolfram to the audacious claim that a mere four lines of computer code are generating reality.

Wolfram admits that his decade of investigation has not yet furnished him with the elusive cosmic computer program – the 'one rule to bind them all'. But he is hopeful that he will one day find it.

One of the most important discoveries to have come out of Wolfram's decade of toil is the recognition that a cellular automaton following rule 110 is far from unique. Wolfram has been surprised to find that many other real systems in the Universe – from turbulent fluids to colliding subatomic particles – also behave as universal computers. In other words, they too have the capacity to simulate any other machine, carry out any conceivable computation.

Because a universal computer can compute, or simulate, absolutely anything, it is trivial to deduce from this that all systems that behave as universal computers can compute as much as each other. In other words, they are equivalent. 'Since universal computers are so widespread in nature, this has far-reaching implications,' says Wolfram. 'It means that everything from the behaviour of a cell to turbulence in a hydrogen cloud drifting in the depths of space to rain pattering on the pavement is equivalent in terms of the computational complexity required to generate it.'

Until now, scientists have assumed that the kind of complexity which is seen in living things – from single cells to human brains – can arise only in a system of large molecules based on carbon atoms. This, after all, is what we observe on Earth. But if, as Wolfram firmly believes, a large range of systems in nature have equivalent computational complexity, it means that the complexity we associate with life is not the unique preserve of planet-bound, water-soluble, carbon-based chemistry. Many of the things we thought were special about life and intelligence can be present in numerous other kinds of physical systems. 'The Universe may contain life forms – including intelligent life forms – the like of which we cannot begin to imagine,' says Wolfram.

He elevates his discovery that large numbers of natural systems have the same computational complexity to an over-arching natural principle. He calls it 'The Principle of Computational Equivalence'. Put crudely, it says that systems of similar complexity are equivalent. Take, for instance, the Earth's atmosphere. According to Wolfram's Principle, because the atmosphere's circulation is as complex as any living thing, it has exactly the same right to be classed as a living thing as you or me! 'People say "The weather has a mind of its own" and think they're just using a metaphor,' says Wolfram. 'I think there's something much more literally true about it.'

Wolfram believes his Principle of Computational Equivalence is a revolutionary and fertile new idea in science. Moreover, he sees it as the next logical step along a road that science first embarked on more than four centuries ago.

In the sixteenth century, the Polish astronomer Nicolaus Copernicus realised that the Sun and planets did not turn about the Earth, as had generally been believed, but that the Earth occupied no special place in the Universe. Later, in the nineteenth century, Charles Darwin deduced that humans were just another product of evolution by the process of natural selection and so they occupied no special place in Creation. Wolfram sees himself as completing the revolution begun by Copernicus and Darwin. There is nothing special, he maintains, about the kind of computation that leads to living things and the thought processes of human brains. Life and intelligence could be implemented in a myriad different physical systems. One consequence of this is that there is no barrier preventing us from creating artificial intelligence – a machine that thinks and behaves like a human being.

All this spells trouble for a kind of reasoning currently favoured by some cosmologists. According to the 'anthropic principle', the reason the Universe has many of the features it has – for instance, laws of physics which permit the formation of galaxies, stars and planets – is because, if it did not, it would not have been possible for human beings to have arisen to notice those features. It is a curiously topsy-turvy logic. And an inevitable consequence is that biology is the ultimate determinant of the physics that we observe around us.

However, the anthropic principle is fatally undermined if, as Wolfram believes, life can be implemented in any number of different physical systems, some as far away from carbon-based chemistry as it is possible to imagine. 'Cosmologists have no right to use the conditions necessary for our existence on Earth to deduce anything about the laws of physics that govern our Universe,' says Wolfram.

Cosmologists wonder why the Universe appears so hospitable for life. The answer, Wolfram believes, is because almost any physical system, almost any set of parameters, can exhibit the complexity of a living thing.

Is God a Programmer?

Everything Wolfram discovered during his decade of toil – the equivalent, he maintains, of hundreds, maybe even thousands, of scientific papers – he eventually distilled into an enormous, epic book. *A New Kind of Science* was finished in January 2002. It was almost 1,200 pages long with about 1,000 black-and-white pictures and half a million words. On the first day of publication it sold 50,000 copies. And it annoyed the hell out of the scientific community.

Absolutely everything about the self-published book seemed to make other scientists see red. Wolfram was accused of not crediting the contributions of others. Wolfram was accused of breathtaking arrogance. After all, he was saying, 'Here in my book is an entirely new way of doing science.' And nobody had dared say that since Isaac Newton.

A striking feature of the venom directed at Wolfram was its swiftness.

Within days of the book's publication, some scientists had posted damning reviews on Amazon's website. Yet the book's 1,200 picture-filled pages were crammed with examples that had to be worked through by the reader. It was hard to believe that anyone could have digested enough to have dismissed it in just a few days.

Chaitin is philosophical about the knee-jerk reaction of the scientific community. 'If you write a book that offends no one and make sure everything you write is absolutely, 100 per cent, correct, then you end up writing nothing,' he says.

One specific criticism is that, although Wolfram has produced a 1,200-page book of pretty pictures of what simple computer programs can do, he has deduced very few universal laws of the kind first discovered by Newton. This, however, is to misunderstand Wolfram. His new kind of science is not at all like the old type – which, of course, is why he has called it 'a new kind of science'. In the old, maths-based science, the motion of, say, a planet travelling around the Sun is distilled into an equation, which predicts its behaviour from now into the infinite past and future. In the new science, the only way to discover how something behaves is to run a computer program. There is no such shortcut. Or, rather, all the shortcuts have already been found – they are conventional, equation-based science.

It is Wolfram's view that much of what is going on in the Universe cannot be distilled into neat equations. You have to run the program to find out what happens. Some of the programs can be run, and a result obtained, more quickly than the Universe. This is because, by some fortunate quirk, some of what the Universe is doing is 'computationally reducible'. 'Almost all of what traditional equation-based science has been doing is looking just at those computationally reducible parts,' says Wolfram.

Wolfram suspects, however, that most of what is going on in the Universe is computationally irreducible. In other words, the only way to find out the outcome of the program the Universe is running is to run it for 13.7 billion years! This raises a spooky possibility. Is the program of the Universe being run by someone or something simply because there is no other way to discover the outcome? In *The Hitchhiker's Guide to the Galaxy*, the Earth turns out to be a computer run by mice to discover the answer to the ultimate question. Might Douglas Adams's jest, by some tremendous irony, actually be near to the truth?

The American physicist Ed Fredkin thinks so. He is convinced that the Universe is nothing more than a computer which is being used to solve a problem. As others have pointed out, this is both good news and bad news. The good news is that there really is a purpose to our lives. The bad news is that purpose may be to help someone or something work out pi to countless zillion decimal places!

The idea that the most fundamental stuff in the Universe – more fundamental even than matter or energy – is information, digital information, is certainly an idea which is taking hold among today's physicists. Those who subscribe to this 'digital philosophy', such as Wolfram and Fredkin, are in absolutely no doubt that what the Universe is doing is computation, in the most general sense of the word. One consequence is unavoidable. Like the insects burrowing in the topsoil of Adams's terrestrial computer, we are a part of the great cosmic computation. 'We never perform a computation ourselves,' says Tomasso Toffoli of Boston University.* 'We just hitch a ride on the great Computation that is going on already.'

---

* Toffoli is famous for inventing the Toffoli Gate, a logical computing circuit which can be implemented by transistors in a computer. Not only is the gate *universal* – which means that any conceivable calculation can be carried out solely by a collection of Toffoli Gates – but it is *reversible* – which means the gate produces the same result regardless of whether current flows through it forwards or backwards. This is important because, in physics, reversible processes use no energy. The Toffoli Gate is therefore extremely energy-efficient.

Of course, if you want to go one mystical step farther and talk about a computation not in its most general sense but in the sense of something directed to some end, like a human computation, then you come to the arena of religious speculation. 'The Universe begins to look more like a great thought than a machine,' wrote the British astronomer Sir James Jeans. And Jeans was really only echoing Bishop George Berkeley, the Irish philosopher who in the eighteenth century declared: 'We exist only in the mind of God.' Chaitin likes to put it in more modern terms: 'Is God a programmer?'

If the idea of the Universe 'computing' something is not mind-blowing enough, consider what it really means if Wolfram is right and the complexity of the Universe is generated merely by applying a simple computer program – a simple rule – over and over again. Information cannot be created out of nothing. Common sense says that what comes out cannot be any more than what is put in. If Wolfram is right, it means that the Universe can contain no more complexity than the simple program responsible for generating it. Consequently, the complexity we see around us cannot be real complexity. It must be 'pseudo complexity'. The Universe only looks complex because we are unaware of the simple underlying rule generating it.

Newton's worldview was one in which the laws of physics orchestrate a predictable world. The planets, for instance, circle the Sun with the regularity of clockwork. However, in such a clockwork universe, where the future is always utterly predictable, scientists faced a conundrum: how can there be any free will?

Wolfram sidesteps this problem. In his clockwork universe, the future of the Universe is predictable – but only in principle. In practice, you can never finish the computations and discover the outcome faster than the Universe does. Free will survives – as pseudo free will!

Chaitin puts Wolfram's worldview in purely mathematical terms. Pi, the ratio of a circle's circumference to its diameter, is a number that appears to be extraordinarily complicated, its digits never repeating, but it can in fact be generated by a short computer program. Chaitin, however, has invented a number which is truly complex. 'Omega' requires an infinitely long computer program to generate it.[*] 'Is the Universe like pi or like Omega?' says Chaitin. 'Most people think it's like Omega. Wolfram thinks it's like pi.'

The reason that most people think the Universe is like 'Omega' – in other words, that it has unadulterated, infinite complexity – is that most people believe in quantum theory. And quantum theory tells us that events in the microscopic world such as the disintegration of an atom or the absorption of a photon of light by a window pane are completely random, as unpredictable as a perfect coin toss. Such events generate an infinite amount of complexity – which is the same as randomness.[*] And this gets permanently imprinted on the Universe – for instance, when a high-energy photon strikes a strand of DNA and causes a mutation, which echoes down the generations, frozen into the fabric of life for all time.

As a consequence of quantum theory, then, much of what we see around us in the Universe is inherently unpredictable. It is the result of countless quantum coin tosses, which have been happening one after the other since the beginning of time. We will therefore never be able to comprehend the Universe in its entirety.

On this score, Wolfram is far more optimistic than the majority of physicists. Because he believes the Universe has finite complexity like pi, he believes that quantum theory as currently practised is wrong. All the randomness that quantum theory generates is therefore really only pseudo randomness, like that in the digits of pi. If he is right, then we may eventually be able to comprehend everything.

[*]   See Chapter 6, 'God's Number'.

Who is right – Wolfram or the rest of the scientific community? Chaitin confesses to spending long hours at Wolfram's house near Boston arguing with him about his ideas. 'In *A New Kind of Science*, Wolfram develops an extremely interesting and provocative vision,' says Chaitin. 'The question is: Does the physical Universe share Wolfram's vision? Time alone will tell.'

3
Yoyo Universe

What happened before the Big Bang? There was another Big Bang and, before that, another . . .

Many and strange are the universes that drift like bubbles in the foam upon
the River of Time.
Arthur C. Clarke, 'The Wall of Darkness', *The Other Side of the Sky*

Everything has been said before but because nobody listens we keep having
to going back and begin all over again.
André Gide

It came out of nowhere like an express train out of the night. Only it
wasn't an express train – it was an entire universe, hurtling towards our
own from a higher dimension. Before the collision, our Universe was an
empty, aching void. In the immediate aftermath, it ignited, exploding
outwards in an unstoppable firestorm of light and matter.

Is this an accurate description of the Big Bang? Is everything we see,
out to the very farthest reaches probed by our biggest telescopes, merely
the wreckage of a titanic collision between universes? A group of
physicists from Britain and America believes the answer is yes. They call
this colliding-universe scenario the 'ekpyrotic universe', from the Greek
for 'born out of fire'. What is more, they say, the cosmic smash-up that
triggered the Big Bang may not have been a unique event. 'Before the

Big Bang, there was another Big Bang and, before that, another one, stretching all the way back through the mists of time,' says Neil Turok of Cambridge University.

A Universe Made Out of String

This awe-inspiring vision has arisen out of 'superstring theory'. Superstring theory, or simply string theory, views the fundamental building blocks out of which everything in the Universe is made not as tiny, point-like 'particles' but as impossibly small 'strings' of super-dense matter. The strings – which are about 10 trillion trillion times smaller than an atom – vibrate exactly like the strings of a violin. And each note they create corresponds to a distinctly different microscopic particle such as an electron or a quark. The higher the pitch of the note, the more energy in the vibration and the heavier the particle.

In the past century, experimental physicists have discovered a host of 'fundamental particles'. These are 'glued' together by four fundamental forces, which in turn are transmitted between the particles by a legion of 'force-carrying' particles.[*] In order to mimic all the myriad properties of such a bewildering zoo of particles, strings must be free to vibrate in a large number of different ways. And this can happen, theorists have discovered, only if strings inhabit a bizarre world with a total of *ten* dimensions of space and time.

---

[*] Nature's four fundamental forces are the electromagnetic force, which glues together the atoms in our bodies; the 'strong' nuclear force and the 'weak' nuclear force, which orchestrate what goes on in the 'atomic nucleus', the tight knot of matter at the heart of an atom; and the gravitational force, which governs the behaviour of planets, stars and the entire Universe.

A ten-dimensional universe is a big embarrassment for physicists. After all, the world about us gives every appearance of having just four dimensions: north–south, east–west, up–down and past–future. However, string theorists refuse to be fazed by the apparent contradiction between their theory and reality. They insist that, hidden from our view, are additional space dimensions. Whereas the familiar space dimensions extend across billions of light years of space and billions of years of time, these hidden dimensions are said to be 'rolled-up' so incredibly small that they have escaped our notice in all experiments to date.

Why go to the trouble of inventing impossible-to-see strings of matter quivering in impossible-to-detect dimensions? The answer, of course, is that there is a big pay-off.

For one thing, string theory resolves a serious conflict between two ideas which are driving forces of modern physics. The first is 'atomism'. This is the belief that, although the world around us looks bewilderingly complex, this is merely an elaborate illusion: beneath the skin of reality there is just a handful of simple, indivisible building blocks. The complexity of the everyday world is nothing more than a manifestation of the enormous number of ways these building blocks can be stuck together. Everything is in the combinations.

At one time the fundamental Lego blocks of reality were thought to be atoms. Currently, they are believed to be even smaller motes of matter called quarks and 'leptons'.

The second important idea driving modern physics is that of 'unification'. This is the belief that many of the fundamental entities we have discovered are really just different faces of entities yet more fundamental.[*] Physicists believe, for instance, that nature's four fundamental forces are merely different aspects of a single 'superforce' and that even

[*] See Chapter 7, 'Patterns in the Void'.

quarks and leptons are different faces of some yet more fundamental particle.

Atomism and unification have proved to be enormously fruitful ideas. But they are set on an inevitable collision course. The reason is that one day, presumably, we will discover the ultimate building block of matter, the indivisible mote out of which everything else is constructed. By definition, it will have no internal structure. After all, if it were constructed from other things, it could be subdivided further and so could not in any way be considered ultimate. But, if it has no internal structure, how can it have different faces? Clearly, it cannot. Like a full stop, it will look the same from every possible viewpoint. Ultimately, then, atomism and unification are irreconcilable.

String theory offers a way out of this impasse. A string is a fundamental, indivisible entity. On the other hand, it can vibrate in a myriad different ways and consequently have innumerable different faces.

Reconciling atomism and unification, however, is not the only pay-off of string theory. Crucially, the theory holds out the tantalising hope of solving one of the greatest outstanding problems in science: how to unite Einstein's general theory of relativity – which describes the force of gravity – with quantum theory, which describes the other three fundamental forces of nature.

Because gravity is a weak force which becomes noticeable only when there is a large amount of gravitating matter about, Einstein's theory predicts the behaviour of big things such as planets circling the Sun and even the whole Universe. Quantum theory's domain of expertise, on the other hand, is the world of small things like atoms and their constituents. Because the domains of the two theories are so completely different, there is generally no overlap, so each theory can be used entirely independently of the other. However, a serious difficulty arises when people want to understand what was going on in the first moments after the Big Bang, the titanic explosion in which the Universe was born about 13.7 billion years ago. At that remote epoch, the Universe was both very massive – the domain of Einstein's theory of gravity – and smaller than an atom – the domain of quantum theory. Neither Einstein's theory nor quantum theory is therefore sufficient on its own to illuminate this remote period. What is desperately needed is a hybrid of the two, an over-arching theory which meshes both together – a 'quantum theory of gravity'.

To say that devising such a theory is hard is a bit of an understatement. Einstein's theory of gravity, in common with all non-quantum, or 'classical', theories, is a recipe for predicting the future. If a planet is here now, the theory predicts that tomorrow it will have moved over there, following a particular path through space. All these things the theory predicts with 100 per cent certainty. Contrast this with quantum theory, which is a recipe for merely predicting possible futures. For an atom flying through space, all that can be known is its 'probable' final position, its 'probable' path. The Herculean task faced by physicists is therefore to unite, or 'unify', two theories – one that deals with certainty and one that deals with uncertainty.

String theory offers hope. It is inherently a 'quantum' theory. And one of the myriad possible string vibrations turns out to have all the properties of a 'graviton', the hypothetical 'carrier' of the gravitational force. Consequently, string theory contains within it a theory of gravity (though not necessarily Einstein's theory of gravity). This is why many physicists see it as the great hope for a unified theory.*

One-dimensional strings turn out not to be the only entities which can pop up in string theory. Because there are ten space-time dimensions to

---

* String theory is not the only hope for uniting quantum theory and Einstein's theory of gravity. There are at least two other promising routes currently being taken by physicists.

play with, the theory can support the existence of more complicated objects, with two, three, four or more dimensions. These are called 'branes'. A string, to use the terminology, is one-brane whereas a more general brane, with p dimensions, is a p-brane – a physicist's rather weak joke.

The existence of branes raises a remarkable possibility. Perhaps that is all our Universe is. Maybe it is a four-brane – a four-dimensional 'island universe' floating in a ten-dimensional space-time. This possibility, in turn, raises another intriguing one. If our Universe is a four-brane, it is unlikely to be the only one. Adrift in the unimaginable ten-dimensional void of string theory there may be other island universes. And, if there are other brane-universes out there, might they occasionally fly close to each other, perhaps even collide?

It is a possibility that has captured the imagination of Turok and his colleague, Paul Steinhardt of Princeton University in New Jersey. 'If there are other universes out there, then it stands to reason that they might occasionally run into each other,' says Turok. 'It might, at long last, explain what the Big Bang was.'

The Train Crash that was the Big Bang

There could of course be countless island universes besides our own lurking out there in the higher-dimensional abyss. However, the simplest scenario is always the easiest to deal with mathematically. Also, nature, for reasons nobody really understands, invariably chooses the simplest option.* Turok, Steinhardt and their colleagues therefore assume that, in

---

* See Chapter 6, 'God's Number', for a discussion of the concrete evidence that this is indeed so. Also, see Chapter 7, 'Patterns in the Void', for a claim that our Universe is simple because it actually contains nothing!

the whole multi-dimensional stringy universe, there are just two lonely four-branes – ours and one other.

Four-dimensional objects are of course impossible for us to visualise since we inhabit a fundamentally three-dimensional world. Faced with this difficulty, Turok and Steinhardt visualise the four-branes as two-dimensional objects – like the two slices of bread in a sandwich. Again, for simplicity, they assume that the two slices of bread are infinite in extent, so that they form the ultimate boundaries of the Universe. They also assume they are utterly empty, without matter or light – and it is impossible to get much simpler than that.

Between the two slices of bread, where the sandwich filling would normally go, is the country of the fifth dimension. And it is along this dimension – which we can no more perceive than a blind man can experience the colour 'blue' – that the two brane-universes hurtle towards each other.

Moving bodies possess energy merely by virtue of their motion. Wander into the path of a speeding cyclist and you will be left in no doubt about this. Consequently, the fast-approaching branes have tremendous 'energy of motion' along the fifth dimension.

As already mentioned, there is a cast-iron rule in physics called the conservation of energy which asserts that energy can never be created or destroyed, only transformed from one form into another. In a light bulb, for instance, electrical energy is converted into an equivalent amount of light energy and heat energy. So, when the branes collide, the conservation of energy ensures that their energy of motion along the fifth dimension is dumped into their four-dimensional interiors as surely as the energy of motion of two colliding express trains is dumped into the twisted wreckage. This energy sets the branes expanding violently. 'It creates the headlong explosion of space we have come to call the Big Bang,' says Turok.

The Big Bang, of course, was a lot more than a violent expansion of completely empty space. We, and the matter out of which we are made, are testimony to that. Here, Einstein has something important to say. In 1905, he stunned physicists by his discovery that mass is actually a form of energy – the most concentrated type of energy of all. Consequently, not only can mass be turned into other forms of energy – for instance, the scorching heat of a nuclear fireball – but other forms of energy can be converted into mass. So, the energy of the colliding branes ends up not only in the furious expansion of empty space but also in the creation of mass – a blistering hot fireball of fundamental particles.

Of course, the same thing happens to the other brane – the one that collides with ours. It too experiences a 'hot' Big Bang.

In this scenario, the subsequent evolution of our Universe is pretty much the same as is widely accepted. As the space of our brane exploded in size, the expansion rapidly cooled the Big Bang fireball. Eventually, when it was cold enough, galaxies and stars congealed out of the shimmering debris. And, one day, after 13.7 billion years, a group of physicists on the third planet of an unremarkable star in a nondescript galaxy called the Milky Way hit on the idea that a collision between peculiar entities called branes might at last provide an explanation for everything we see around us.

How the Universe Got its 'Spots'

Explaining everything we see around us, however, means explaining a lot more than simply the explosion of the blisteringly hot matter of the Big Bang. There is the small matter of how the Universe got to look the way it does today.

Early on in the history of our Universe, matter was spread extremely smoothly throughout space. We know this from the heat of the Big Bang fireball, which still permeates every pore of space. After all, it was bottled up in the Universe and had nowhere else to go. Greatly cooled by the expansion of space in the past 13.7 billion years, this 'afterglow of creation' appears today not as light visible to the naked eye but as invisible microwaves of the kind used by mobile phones and microwave ovens.* One of the most striking features of this 'cosmic background radiation' is its smoothness: it arrives at Earth equally from all directions. This tells us something important about the way matter was distributed shortly after the Big Bang. Since it was mixed in with the heat radiation in the Big Bang fireball, it too must have been smeared remarkably smoothly throughout space.

Yet, today, far from being spread evenly throughout space, the matter of the Universe is tied up in 'galaxies', great islands of stars, separated by great voids of empty space. One of the key questions in cosmology is: How did the Universe go from being smooth to being lumpy?

A partial answer was provided by NASA's Cosmic Background Explorer (COBE) satellite. In 1992, it discovered that, although the afterglow of the Big Bang looks perfectly smooth around the sky, in some directions it is ever-so-slightly brighter and in others ever-so-slightly fainter than in others. These slight irregularities in the cosmic background radiation are believed to mark the embryonic 'seeds' of great clusters of galaxies in today's Universe. They mark denser-than-average regions of the early Universe, which, by virtue of their stronger-than-average gravity, were able, as they grew older, to pull in more matter than neighbouring regions. This increased their gravity, enabling them to suck in yet more matter, and so on – a cosmic instance of the rich getting ever richer.

* See my book, *Afterglow of Creation* (University Science Books, Sausalito, California, 1996).

The cosmic background radiation in fact comes to us from an epoch when the Universe was already about 450,000 years old. Since the seeds of galaxy clusters in today's Universe were already present at that time, it follows that they must have been imprinted on the Universe at an even earlier time. But by what? Turok and Steinhardt's colliding-universe scenario appears to provide an answer. Everything hinges on the restless churning of the 'vacuum'.

Usually, we think of the vacuum of space as completely empty. This is not, however, the way modern physics sees it. As already pointed out, quantum theory permits energy to pop into existence at any time out of absolutely nowhere. The proviso is that, within a split-second, it must pop back out of existence. Think again of that teenager who borrows his dad's car for the night but gets it back in the garage before his dad gets up the next morning and notices its absence. Well, energy is a bit like the borrowed car. In the microscopic world the law of conservation of energy is not such a cast-iron rule. It fails to notice energy popping into existence out of nothing as long as it pops back out of existence quickly enough.

Now, energy, according to Einstein's general theory of relativity, warps the space-time around it. Usually, of course, we think of gravity as a property of mass but, since mass is a form of energy, this is in perfect accord with general relativity. The upshot is that the energy which is permitted by quantum theory to pop briefly into existence actually warps the space-time around it. If we could examine empty 'space' extremely closely with some kind of 'super-microscope', we would not see it as smooth and unruffled. Instead, the 'quantum vacuum' would be in a state of ceaseless convulsion, seething and churning like a saucepan of boiling water.

This has profound implications for two branes in the last moments before their titanic Big-Bang-triggering collision. The reason is that the branes themselves have enormously powerful gravity. And, when the two slices of bread come so close there is hardly any gap between them, the gravity of one brane plucks the surface of the other. This has the effect of making the churned-up surface of each brane even more churned up.

Look at it another way. Imagine you can view the surface of our brane with that super-microscope. Furthermore, imagine you can freeze the image for an instant. What you will see is a terrain reminiscent of the Himalayas. Clearly, the gravity of the approaching brane has the greatest effect on the tallest peaks since they are closer to its influence. It is these peaks that, like soft toffee, are pulled even higher at the expense of their less lofty counterparts.

This mechanism, according to Turok and Steinhardt, greatly magnified the lumps and bumps on our brane. They were the ultimate 'seeds' of clusters of galaxies in today's Universe. Once the branes collided and matter was conjured into being to fill our brane, that matter would have naturally gravitated towards them. After all, the more warped a region's space-time the stronger its gravity. The two things are one and the same. If Turok and Steinhardt are right, then the biggest structures in existence in today's Universe were spawned by regions of space-time smaller than an atom.

How the Universe Got to be So Smooth

But explaining how the matter of our Universe went from being smooth to being lumpy is only one difficulty that must be addressed by a theory of the origin of the Universe. Another is how today's Universe is so smooth.

But – wait a minute – we've agreed that it isn't smooth, it's lumpy. Well, that's perfectly true if you look at individual galaxies and clusters of galaxies. However, if you look across great swathes of space, averaging out all the irregularities, it turns out that the density of matter is remarkably similar from place to place. What is more, the cosmic background radiation is extraordinarily uniform as well – apart, of course, from those very tiny irregularities found by COBE.

As already pointed out, all this poses a problem for the standard picture of the Big Bang. As the Universe expanded and cooled, inevitably the temperature of some bits fell a bit faster than others, the density of some bits dropped a bit faster than others. Since heat flows from hot regions to cold regions, and matter flows from a dense region of gas to a less dense region, these temperature variations and density variations might be expected to equalise out. However, there is a problem.

As noted before, the cosmic background radiation comes from an epoch 450,000 years after the moment of creation. Light at that time could have traversed a maximum of 450,000 light years. But, if you imagine the expansion of the Universe running backwards, like a movie in reverse, you discover that 450,000 years after the moment of creation the entire observable Universe was something like 18 million light years across.

According to Einstein, nothing can travel faster than light. It is the ultimate speed limit. So, as the Universe expanded and cooled, heat could not have spread fast enough from hot regions to cooler regions; matter could not have spread fast enough from dense regions to less dense regions. No influence could have travelled more than 450,000 light years, a tiny fraction of the 18 million light years from one side of the Universe to the other. There was simply insufficient time for the temperature and density of the Universe to equalise.

The unavoidable conclusion is that today's Universe cannot possibly have the same matter density everywhere; neither can the temperature of the cosmic background radiation be the same everywhere. The trouble is they *are*. The standard Big Bang theory is therefore in serious conflict with our most basic observations of the Universe. To fix it, something else is needed. A missing ingredient.

As pointed out before, the missing ingredient, according to consensus opinion, is 'inflation'. According to the theory, the Universe, in its first split-second, underwent an ultra-brief phase of super-fast expansion.

No one knows the fine details of inflation, only that it could have been driven by a bizarre state of the quantum vacuum – one in which gravity blew rather than sucked. The key thing is that, if the Universe did indeed undergo a phase of super-fast expansion, the entire observable Universe could have come from a far smaller volume of space than we would naively expect simply from running the movie of the expansion backwards. And, if it came from a far smaller volume, then heat and matter would have had plenty of time to equalise the temperature and density. The theory once again accords with our observations of the Universe. Inflation has the added advantage that it 'inflates', or magnifies, the roiling quantum vacuum – including those seeds of cosmic structure – providing a convenient explanation for the origin of galaxies in the Universe.

A new and elegant way of keeping the Universe uniform, however, is suggested by Turok and Steinhardt's colliding universe scenario. On the microscopic scale, the approaching branes are like a choppy sea. However, on a larger scale these contortions average out, so that the branes are to all intents and purposes utterly flat. What this means, is that as the branes first touch, they touch everywhere at once. As the energy of motion of the branes is converted into particles and heat, it is everywhere in the Universe converted into exactly the same density of particles at exactly the same temperature. Hey presto, the uniform density and temperature of today's Universe is explained.

So far, so good, then. The brane-collision scenario can explain what the Big Bang was. It can explain where the galaxies in today's Universe came from. And it can explain why matter is spread so smoothly throughout space and why the afterglow of the Big Bang is exactly the same everywhere in the Universe. But that still leaves the mystery of why, 13.7 billion years ago, two branes sailing through the ten-dimensional void just happened to slam into each other.

To have generated such a violent collision, some force must have snapped the branes together. Since the only thing between the branes was the vacuum, the only possible culprit is the vacuum. Like a *Star Trek* tractor beam, it must have exerted an irresistible force of attraction on the branes.

If the idea of the vacuum exerting a force sounds preposterous, think again. Not only can the vacuum do such a thing but exactly this kind of vacuum-generated force turns out to be the dominant influence controlling the evolution of our Universe today!

The Spring in the Fifth Dimension

Evidence for such a vacuum force first came to light in 1998. Two teams of researchers were observing 'supernovae' in distant galaxies. One team was led by the American Saul Perlmutter and the other by the Australians Nick Suntzeff and Brian Schmidt. Supernovae are exploding stars which often outshine their parent galaxy and so can be seen at great distances out in the Universe. The kind the two teams were looking at were known as 'Type Ia supernovae'. They have the property that, when they detonate, they always shine with pretty much the same peak luminosity. So, if you see one that is fainter than another, you know it is farther away.

What the astronomers saw, however, was that the ones that were farther away were fainter than they ought to be, taking into account their apparent distance from the Earth. The only way to explain what they were seeing was that the Universe's expansion had speeded up since the stars exploded, pushing them farther away than expected and making them appear fainter.

But, as noted earlier, this was contrary to all expectations. In the aftermath of the Big Bang, the galaxies – the building blocks of the Universe – are flying apart from each other like pieces of cosmic shrapnel. The only force acting on them should be their mutual gravity and this ought to be pulling them together and 'braking' the expansion of the Universe.

The discovery that the expansion of the Universe is in fact speeding up implied that there is a hitherto unsuspected force operating in the Universe. It is overwhelming gravity and driving the galaxies apart. Its origin can only be in the space between galaxies. Far from being empty, the vacuum must be filled with some kind of weird anti-gravity stuff. It is this 'dark energy', mentioned earlier, that is remorselessly driving the galaxies apart.

The first thing to say about dark energy is that nobody understands it. In fact, it is a huge embarrassment. Quantum theory – the very best theory physicists possess – predicts that, if a chunk of vacuum has any energy at all, it should have 1 followed by 123 zeroes more than is observed! This has been described by the Nobel Prize-winner Steven Weinberg of the University of Texas as: the 'worst failure of an order-of-magnitude estimate in the history of science'.

Though quantum theory cannot explain the amount of energy

observed to be in the vacuum, Einstein's theory of gravity nevertheless provides a means by which the energy of the vacuum can exert a force. It is all down to the 'source' of gravity. Newton thought it was mass; Einstein realised it was any form of energy, which includes mass–energy. This, however, is not the complete story. As pointed out before, a close inspection of Einstein's 'field equations' of gravity reveals that the source of gravity is in fact energy – strictly speaking, energy density, the energy in a unit volume of space – plus three times the pressure. It's this second mathematical term that makes all the difference.

In all normal circumstances, the energy density of matter far exceeds any pressure it exerts on any container in which it is confined, making the pressure term of no consequence whatsoever. The possibility nevertheless exists that our Universe contains a material totally unlike anything we know of – a material with a pressure comparable to its energy density. If the pressure is positive – that is, it pushes outwards like the helium gas in a Mickey Mouse balloon – then its effect is to increase the material's gravity. If the pressure is negative – that is, it pulls things inwards like the tension in a piece of stretched elastic – then, incredibly, the material's gravity can in fact reverse.* As already mentioned, gravity can blow instead of suck.

Here then is the recipe for the invisible dark energy which fills our Universe. Space is evidently filled with invisible stuff the like of which we have never before imagined. Everywhere it is trying desperately to shrink. Yet, paradoxically, it is causing the Universe to expand ever faster.

The discovery of the key importance of the vacuum in our Universe has come as a great shock to the scientific community. Nobody – least of all, the physicists who are at a loss to find a plausible explanation –

---

* To be precise, this requires the material's pressure to be less than -⅓ its energy density.

wanted it. To Turok and Steinhardt, however, the vacuum is not some unwelcome entity that throws a spanner in the cosmic works. Far from it. In the colliding-universe scenario, it is an absolutely essential ingredient. It is the vacuum that creates the force of attraction which pulls the branes together in the fifth dimension. It is the vacuum that smacks them together like clashing cymbals to create the Big Bang.

But surely the vacuum in today's Universe is pushing everything apart, not pulling things together? True enough. Recall, however, that the vacuum has the ability both to blow and to suck – it all depends on how its pressure is related to its energy density. Well, remarkably, it is possible for the vacuum to suck in the fifth dimension while simultaneously blowing within the four-dimensional island of our brane-universe. In one stroke, the vacuum can explain both the force that dragged the branes together and the force which is currently causing the expansion of our Universe to speed up.

But the vacuum can do even more than this. The relationship between its pressure and its energy density need not stay the same for all time. It can change. The vacuum can go from blowing to sucking and vice versa. And this has profound implications for Turok and Steinhardt's brane-collision scenario. It provides a means by which Big-Bang-generating collisions can happen over and over again.

When the branes are far apart, the vacuum in the fifth dimension sucks, pulling them together. The branes eventually collide and actually pass right through each other. Now, however, when the branes are close together, the vacuum changes from sucking to blowing. It drives the branes away from each other until, when they are far apart again, the vacuum changes from blowing to sucking and the whole process repeats itself.

In effect, the vacuum, which blows continually in our four-dimensional Universe, acts like a spring between the branes in the fifth dimension. When the branes are apart, the spring is stretched and under tension, and pulls the branes together. When the branes are close, the spring is squeezed and compressed, and pushes the branes apart. In this way, the branes come together and collide, move apart, then come together and collide again, over and over again.

If Turok and Steinhardt are right, there was not just one Big Bang but a whole series of Big Bangs, stretching back into the infinite past. This in itself is not a new idea. In its previous incarnation, people called it the 'oscillating', or simply 'bouncing', universe.

The Bouncing Universe

In the bouncing universe, the gravity of all the galaxies tugging on each other eventually slows their fleeing motion to a standstill. The expansion having run out of steam, the Universe embarks on a phase of runaway contractions. In fact, it shrinks all the way down to a 'Big Crunch' – a sort of mirror image of the Big Bang in which all matter is crammed into the tiniest of tiny volumes. But this is not the end of the story. Matter is hypothesised to have some residual stiffness, like a hard rubber ball which can be squeezed so far but no farther. At the very last moment, therefore, the Universe rebounds in another Big Bang, which is followed by another Big Crunch and another Big Bang, and so on, ad infinitum.

The idea of such a bouncing universe was once very popular with cosmologists because it provided an answer to the perennially awkward question: what happened before the Big Bang? According to the bouncing-universe scenario, before the Big Bang there was an earlier Big Bang. And, before that, an even earlier one. Far from being a unique event, our Big Bang was merely one Big Bang in a never-ending cycle of Big Bangs and Big Crunches. Despite the undoubted aesthetic appeal of the bouncing universe, however, the theory was scuppered by several serious difficulties.

One arises because, in each Big Bang–Big Crunch cycle, new stars are born. Because these pump heat into space, they make the next Big Bang hotter than its predecessor. And, the hotter the Big Bang, the bigger the Universe grows before its expansion runs out of steam. This may not seem a problem. However, imagine looking backwards in time like some omniscient god. The Big Bang–Big Crunch cycles get progressively cooler and smaller in amplitude until, eventually, they dwindle away to nothing. This is the moment at which they began.

But if the cycles had a beginning – no matter how far back in the ultra-remote past that was – all we have really succeeded in doing is replacing the awkward question – What happened before the Big Bang? – with the equally awkward question – What happened before the first Big Bang? This hardly counts as progress.

An even more serious flaw in the bouncing-universe scenario was found in the 1960s by Stephen Hawking and Roger Penrose at Cambridge University. They proved that, if Einstein's theory of gravity provides the correct description of the Universe, then the Universe must have begun in a 'singularity'. This is a region of space where the density, the temperature, and so on, all sky-rocket to infinity. The appearance of a singularity in any theory of physics signals the total breakdown of predictability. It is therefore impossible to determine what happened before the singularity. The very concept of 'before' has no meaning. In effect, Hawking and Penrose had dropped an opaque curtain across the Big Bang, obscuring for ever the view of earlier times.

If general relativity could not say anything sensible about the period 'before the Big Bang', the bouncing universe, with its endless pre-Big Bang cycles, was clearly a non-starter. How, then, at the beginning of the twenty-first century, is it possible for Turok and Steinhardt to revive a cosmological scenario with multiple Big Bangs stretching back into the infinite past? They can do it, it turns out, because their scenario is only superficially similar to the bouncing universe. In fact, it has major differences. And it is these differences that ... well ... make all the difference.

The Cyclic Universe

To emphasise that the brane-collision scenario is not the same as the bouncing universe, Turok and Steinhardt have christened it the 'cyclic universe'. One profound difference is that it arises not out of Einstein's theory of gravity but out of string theory, which proponents believe is a more fundamental and accurate description of reality. Since the 'singularity theorems' of Hawking and Penrose apply only to Einstein's theory, Turok and Steinhardt claim that the Universe never goes through a singularity in the cyclic scenario.

From the point of view of general relativity, of course, there is a singularity at the moment the branes touch. This is the Big Bang. However, Turok and Steinhardt maintain that, from the point of view of string theory, the 'scale factor', which sets the size of space-time on a brane, remains perfectly finite. In other words, at the very moment Einstein's theory says the matter on the branes should have a density and temperature which is sky-rocketing to infinity, string theory declares that everything is in fact quite well-behaved.

How can general relativity and string theory look at exactly the same event and one see a singularity and one not? Turok believes the answer is that the singularity here is not actually a real thing. It is merely an artefact of our point of view. 'Look at the Big Bang the wrong way – from the point of view of general relativity – and you get a singularity,' he says. 'Look at it the right way – from the perspective of string theory – and there is no singularity.'

There is a precedent. Another singularity in the theory also turned out to be an artefact of our viewpoint. A year after general relativity's birth in 1915, the German physicist Karl Schwarzschild was serving in the trenches of the First World War when he discovered within general relativity a description of a non-spinning black hole.[*] The 'Schwarzschild solution' sported a singularity at the 'event horizon', the imaginary membrane surrounding a black hole which marks the point of no return for matter spiralling in. This singularity, however, turned out to be simply an artefact of the particular 'coordinate system' which Schwarzschild had used. 'When he changed to another system, the singularity went away,' says Turok. 'The event horizon of a black hole is not a place of infinite density. For a big enough black hole, it can even be crossed in safety.'

In the same way, Turok and Steinhardt hope that the singularity in the brane-collision scenario is a mirage which goes away when looked at through the spectacles of string theory. 'String theorists are currently working to see if they can prove it,' says Turok. 'We are hopeful.'

However, the problem of the singularity was not the only problem to beset the bouncing universe. There was the awkward fact that the bounces got bigger and bigger with time, implying that the Big Bang-Big Crunch cycles could not have been going for ever but must have begun at some moment in the remote past. Such a 'beginning' to the Universe is aesthetically unsatisfying. It can be avoided in a bounce scenario only if each cycle remains the same size as its predecessor. This, it turns out, is exactly what happens in the cyclic universe.

[*] From the point of view of Einstein's theory of gravity, a black hole is a region of space so grossly warped that it can be imagined as a bottomless well. Not surprisingly, nothing, not even light, can clamber out.

In the bouncing universe, each Big Bang is bigger than its predecessor because the stars which form in each cycle pump out heat into space, making succeeding bangs hotter. However, in the brane-collision scenario this does not happen for one very good reason – the presence of the vacuum.

The vacuum both sucks and blows along the direction of the fifth dimension, repeatedly pushing the branes apart and pulling them back together. However, this is not what happens within the space of each brane. Here the vacuum does nothing but blow. It is blowing today, as Perlmutter and others discovered in 1998. It is causing the space of our brane-universe to expand ever faster. And it is precisely this accelerated expansion of space that rescues the cyclic universe from the same fate as the bouncing universe.

An unavoidable consequence of this runaway expansion is that all the matter in the Universe – and of course all the heat pumped into space by stars – will eventually be smeared incredibly thinly throughout a tremendous volume of space. In effect, all the stuff in the Universe will be diluted out of existence. The brane will be returned to essentially the state it was in when the brane collision triggered a Big Bang. Consequently, when the next brane collision triggers the next Big Bang, that bang will be precisely as big as its predecessor.

In the cyclic universe, therefore, it really is possible to have a never-ending series of Big Bangs marching backwards into the infinite past and forwards into the infinite future. The awkward question of what happened before the Big Bang finally has an answer. It is Big Bangs all the way back!

Of course, even if the question of what happened before the Big Bang has an answer, there still remains the even more awkward question – Why a Universe with an infinite series of Big Bangs rather than something else? Or, more succinctly, why is there something rather than nothing? This is the grandaddy of all cosmological questions. It may even be beyond the capability of science to answer . . . *

An Endless Series of Big Bangs

The bouncing universe not only had Big Bangs but Big Bangs alternating with Big Crunches. And this highlights a major qualitative difference between the bouncing and cyclic scenarios. In the cyclic universe, the oscillations occur in the invisible, and unobservable, fifth dimension, with the branes coming together, colliding, flying apart, then coming together again. There is no oscillation in our four-dimensional Universe. Instead, as the branes collide and re-collide, the space of our brane-universe is repeatedly subjected to bursts of headlong expansion. As each runs out of steam, there is a new bang which starts everything expanding again.

One instant our brane-universe is empty. The next there is a Big Bang and all of space is filled with super-heated matter and light which expands explosively.† The expansion of space eventually cools the matter enough that it can congeal into galaxies and stars and planets. Eventually, after tens, maybe hundreds of billions, of years, the expansion of space dilutes matter and light effectively out of existence. Our brane-universe is empty. The cosmic slate is wiped clean. Suddenly, there is another Big

---

*   Or maybe not. See Chapter 7, 'Patterns in the Void'.
†   A central feature of the Big Bang, which is often difficult to comprehend, is that it happened everywhere at once. It was not an explosion located at a single place like the explosion of a stick of dynamite. Rather, all of space was filled with light and matter and began expanding everywhere simultaneously. Astronomers often use the crude analogy of a rising raisin cake. Every raisin recedes from every other raisin. None is at the centre of the expansion. So it is with galaxies in our Universe.

Bang and all of space is filled with super-heated matter and light which expands explosively . . . In this way, our brane-universe expands for ever, the expansion of space boosted periodically by brane collisions in a hidden fifth dimension.

'Some say the world will end in fire, Some say in ice,' wrote the poet Robert Frost. In the cyclic universe, the Universe alternates for ever between phases of ice and fire.

In all this, the vacuum is of crucial importance. Not only is it responsible for repeatedly crashing together the branes like cosmic cymbals in the fifth dimension but it is responsible for the enormous expansion of space necessary to dilute starlight and ensure that Big Bangs do not get bigger with time. By contrast, in the standard picture of cosmology – which consists of the Big Bang plus inflation – the vacuum which is currently speeding up the expansion of the Universe is simply an arbitrary phenomenon which must be bolted onto the model in order that it should accord with reality. What is more, the vacuum has to have two states with vastly different energies – the ultra-low-energy state governing the expansion of today's Universe and the ultra-high-energy state which drove inflation during the first split-second of the Universe's existence.

Endless Repetition, Endless Novelty

But, isn't the cyclic universe, with cycles all identical, its endless repetition, mind-numbingly dull? Turok and Steinhardt think not. 'Just because the cycles repeat does not mean the events in each cycle are identical,' says Turok. 'The laws of quantum mechanics, which govern the microscopic world of atoms and their constituents, are inherently unpredictable – they exhibit randomness,' says Turok. 'Consequently, the detailed sequence of events in each cycle is different.'

More speculatively, Turok points out that, though our four-dimensional brane-universe would behave the same from one cycle to the next, the extra rolled-up dimensions might vary their sizes. The significance of this is that the fundamental forces of nature are suspected to be merely manifestations of these hidden dimensions. 'The laws of physics could actually change from cycle to cycle,' says Turok.

This might at last explain a peculiar, and much-remarked-upon, feature of our Universe. The laws of physics appear to be 'fine-tuned' for the existence of stars, galaxies and life. For instance, if the force of gravity were only a few per cent stronger than it is, it would crush and heat up the cores of stars to such an extent that they would burn their hydrogen fuel in less than a billion years. This would be an insufficient time for the evolution of complex life, which on Earth has taken the best part of four billion years. On the other hand, if the force of gravity were only a few per cent weaker than it is, it would be unable to crush and heat up stellar cores sufficiently to even burn their hydrogen fuel. Stars like the Sun would be an impossibility.

This is but one example of the fine-tuning of the laws of physics. If the strength of any of the other fundamental forces, or the masses of fundamental particles, were even slightly different, there would be no stars or planets or life.

There would appear to be only two logical explanations for the fine-tuning we observe in our Universe. Either 'God' fine-tuned the Universe for life – though perfectly acceptable, this has the drawback that it prevents any further scientific enquiry. The other possibility is that there is more than one universe – in fact, a huge number of universes – each with different laws of physics. Among this vast 'multiverse' of universes are one or more with the laws of physics necessary for the creation of galaxies and stars and life. We have obviously arisen in such a universe, goes the argument. After all, how could we have arisen in any other?

If there is a multiverse, an obvious question raises its head: where are the other universes? Turok and Steinhardt's scenario suggests an answer: in the other cycles. In other words, the laws of physics are different in every cycle and – surprise, surprise – we have arisen in the only cycle where it was possible for us to arise.

Ripples in the Fabric of Space

Turok admits to being very excited by all the possibilities of the cyclic universe. 'The whole thing has just fallen into our laps,' he says. 'We didn't set out to resuscitate the bouncing universe, yet so many things all appear to be slotting into place.'

'It's taken my breath away,' says Steinhardt. 'I have been both shocked and elated at how we have proceeded from a vague, intuitive notion and constructed a model as compelling and powerful as the cyclic universe.'

So much for the cyclic universe's advantages – aesthetic and otherwise – does it represent reality? The only way to tell is if there is some measurable property of our Universe for which the cyclic scenario makes a prediction at odds with the standard cosmological model – the Big Bang plus inflation. It turns out there is.

According to Einstein's theory of gravity, whenever matter is moved violently, it produces gravitational waves – actual ripples in the fabric of space-time which propagate outward from their source in much the same way that concentric ripples spread out from an impacting raindrop on the surface of a pond. Inflation, the super-fast expansion of space, would have involved just about the most violent movement of matter imaginable. So, if it did really happen during the first split-second of the Universe's existence, it should have generated copious gravitational waves. They should survive in today's Universe as a chaotic 'background' of space-time ripples. Crucially, however, no such gravitational waves are generated in the cyclic universe.

Here then is a critical test. If there is no background of gravitational waves, the cyclic universe is correct. If there is, it is wrong. Simple. Except that gravitational waves are extremely weak and nobody has yet succeeded in detecting them on Earth.

Nevertheless, the hypothetical gravitational waves of inflation should affect the way the temperature of the cosmic background radiation varies from place to place in the sky. It is a very subtle effect but it might just be detectable by the European Space Agency's 'Planck' space probe, due for launch in 2007. If so, we should soon know whether our universe did indeed begin in the ultimate train crash – a collision between island universes.

Part Two
THE NATURE OF REALITY

4
Keeping It Real

Why do we never see the weird world that underpins the everyday world? Because we never observe it – we only ever observe ourselves!

We have found a strange footprint on the shores of the unknown. We have
devised profound theories, one after another, to account for its origins. At last,
we have succeeded in reconstructing the creature that made the footprint.
And lo! It is our own.
Arthur Eddington, *Space, Time and Gravitation*, 1920

We don't see things as they are, we see things the way we are.
David Mitchell, *Number9dream*, 2001

The football sits on the penalty spot. The crowd in the stadium is
hushed. The player with the hopes of his team mates on his shoulders
looks up, takes a deep breath and runs towards the spot. His boot
connects with the football. It loops to the left. It loops to the right. It
scythes along the ground. In fact, it flies along a thousand different
trajectories as if a thousand different penalty takers have struck a
thousand separate footballs. The kicker jumps with joy, the kicker drops
to his knees in despair; the ball thwacks into the back of the net, the
goalkeeper palms it safely away; the crowd roars, the crowd groans.

What is going on? This is not reality. No, it is not. But the question is,
why not? It sounds like a totally ridiculous thing to ask. But, actually, it
is entirely sensible. In fact, it is just about the most profound question
that can be asked of reality.

You see, everything, including a football, is made of atoms. They are the Lego bricks from which the world is assembled. And it has long been known that an atom flying through space really does fly along multiple trajectories simultaneously (it sounds unbelievable – but there you are). So, when a football is kicked, it ought to follow multiple paths through the air like the atoms from which it is made. But it does not. Why? Or, to put it another way, why is the world of atoms – the world that underpins our world – so different from the world of footballs and people and planets? In short, where does the everyday world come from?

Remarkably, it is only in the last decade or so that physicists have zeroed in on the answer to this fundamental question. And the answer, it turns out, is both surprising and deeply, deeply subtle.

Waving Goodbye to Certainty

To understand how it is that an atom can do many things at once – like fly along multiple trajectories through space – you first need to know some pertinent history. In the course of the past century, a multitude of laboratory experiments have shown that atoms and their 'subatomic' constituents, such as electrons, have a peculiar, schizophrenic nature. They behave not only as 'particles' – localised objects like microscopic billiard balls – but also as 'waves' – smeared-out entities like ripples on a pond. The twist is that the wave associated with an atom is not a tangible thing like a wave on the surface of a body of water; it is an abstract, mathematical thing. Nevertheless, it can still be imagined spreading through space just like a real wave. There is even an equation which predicts exactly how the wave propagates. It is called the Schrödinger equation, after the Austrian physicist Erwin Schrödinger.

Since the wave associated with an atom is an abstract thing, it might be expected that it is of no practical consequence. Nothing could be farther from the truth. The wave does indeed make contact with reality. It does it through its height, or 'amplitude', which the Schrödinger equation allows to be calculated at any location in space. In fact, the important thing is the 'square' of the wave's amplitude. This quantity turns out to be the chance, or 'probability', that the atom will actually be found at the location if anyone cares to take a look. For instance, the probability of finding it at one location might be 25 per cent and 5 per cent somewhere else.

Already, this highlights a dramatic difference between the large-scale, everyday world and the small-scale world of atoms. Whereas you can know with 100 per cent certainty where, for instance, your car is parked – unless it has been towed away, it is around the corner where you parked it last night – prior to looking, we can never be exactly sure where an atom is located. All we can know is that it has a certain probability of being over here, a certain probability of being over there, and so on (the moment an atom is 'observed', however, it is pinned down to one place, and one place only, and all the possibilities that existed prior to the observation cease to exist).

Not knowing where an atom is with certainty turns out to be only one of the many peculiarities of the Alice-in-Wonderland world of atoms. Not only is it impossible to know with absolute certainty where an atom is located, it is impossible to know with certainty what it is doing. There is no way, for instance, to know which path an atom will take as it flies through space – only that it will take one path with a particular probability, another path with another probability, and so on. The fact that atoms can behave like waves as well as particles also has other profound consequences for the microscopic world. These follow unavoidably from the fact that everything that real waves can do, so too can the abstract waves associated with atoms. Take the tendency for waves on water to spread out with time. The abstract waves associated with atoms do the same thing. It means that, in the microscopic world, the longer you wait, the greater is your uncertainty about where an atom actually is. But the property of real waves which has the most shocking ramifications in the microscopic world is a less well-known one – their ability to combine with each other to form composite waves.

Two Places at Once

Anyone who has watched the sea knows that you can get big, rolling waves, and you can get small ripples, caused by the breeze. They will also know that you can get a combination of the two – a big, rolling, wave with small ripples superimposed on it. This turns out to be a general property of waves of all kinds. If two or more types of wave are possible, then a combination, or 'superposition', of the waves is also possible.

In the everyday world, this is a mundane observation. However, in the world of atoms and their constituents – the quantum world – it has consequences which are . . . well . . . earth-shattering. Consider, for instance, a quantum wave that is big in one particular location – that is, there is a high chance of finding the atom there if you look. Now, consider a quantum wave that is big in some other place. In other words, the atom has a high probability of being found in the other place. Well, if these two waves are possible, then it follows that a combination is also possible. Nothing too remarkable about this, you might think. Until it dawns on you what such a superposition in fact represents: the atom literally being in two places at once – the equivalent of you being in New York and London at same time!

This is no theoretical fantasy. It is actually possible to observe an atom being in two places at once – or, to be more precise, the consequences of it being in two places at once. There is a famous physics experiment in which atoms, or other microscopic particles, are fired like tiny bullets at an opaque screen with two close-together vertical slits cut in it. In the 'double slit' experiment, the quantum waves representing the atoms go through the slits and mingle with each other on the far side of the screen. Where the crests of the two waves coincide, the waves are boosted, and where the crests of one coincide with troughs of the other, the waves cancel each other out. This reinforcing and cancelling phenomenon is known as 'interference' and is common to all types of waves.

Because of interference, when a second screen is placed in the path of the mingling atoms, there will be places on the screen where the quantum wave associated with them is big and places where the quantum wave is non-existent. These will correspond to places where lots of atoms hit the second screen and places where no atoms at all hit the screen. If an atom makes some kind of black mark where it hits the screen, the result will be a pattern of vertical black-and-white stripes not unlike a supermarket bar code.

The crucial prerequisite for such an 'interference' pattern is that two things mingle – in this case, the waves emerging from one slit with the waves emerging from the other. Remarkably, however, the pattern on the second screen still builds up even if the atoms are fired at the first screen one at a time, with long intervals of time in between. The unavoidable conclusion is that each atom mingles with itself. In other words, it goes through both slits simultaneously – it is in two places at once.

And, if being in two places at once is not bad enough, there is worse. The quantum wave associated with an atom represents more than simply the probable location of the atom. It represents everything that nature permits us to know about the atom such as its speed, its energy, how it might be spinning and so on. Consequently, the fact that a superposition of quantum waves is possible does not simply mean that an atom can be in two places at once. It also means it can do two things at once – the equivalent of you walking the dog and washing the car at the same time.

If things in the everyday, large-scale world behaved like microscopic, quantum particles, a skier, faced with a tree blocking their path, would go both ways round it at once. And, as pointed out, a football struck from the penalty spot would fly through the air along all conceivable paths.

So, how is it that atoms and their like can be in many places at once and do many things at once whereas skiers and footballs – which are merely large assemblages of atoms – cannot? Why do we not ever see a person walking through two doors simultaneously? Or witness a superposition between a giraffe and a zebra?

Quantum Theory Applies Except When It Doesn't

For a long time, the standard explanation of why we do not see weird superpositions was provided by the so-called Copenhagen interpretation of quantum theory. This says that, when an atom is not being 'observed', its associated quantum wave spreads through space according to the edict of the Schrödinger equation. The atom has a certain probability of being over here, a certain probability of being over there, and so on. As soon as the atom is observed, however, things change abruptly. Somehow – and nobody knows quite how – the 'act of observation' forces the atom to stop misbehaving and plump for being in one location with 100 per cent probability. In the jargon, it 'collapses' the quantum wave down to a single possibility, in the process destroying all other possibilities.

In essence, the Copenhagen interpretation says that quantum theory applies except where it does not. And where it does not is when an 'observation' is made.

To say the Copenhagen interpretation is unsatisfactory is a bit of an understatement. For a start, the collapse of the quantum wave is a phenomenon tantamount to magic. Physicists put it into the theory in an ad hoc way and it appears to happen instantaneously – that is, in no time at all. This is of course ridiculous. Nothing in the real world happens in no time. As quantum physicist Anton Zeilinger of the University of Vienna says: 'The speed of collapse is bull.'

Even worse, the Copenhagen interpretation does not even specify what exactly constitutes an 'observation'. Does an atom have to be observed by a mindless particle detector or by a conscious human being? If the latter is the case, then it could be said that the everyday world of certainty does not exist without conscious observers to observe it. 'Does the Moon exist if nobody looks at it?' asked Einstein. If you believe the most extreme form of the Copenhagen interpretation, the answer is no! Which is why Einstein asked the question – he hated the Copenhagen interpretation of quantum theory and missed no opportunity to highlight its inadequacies.

A less extreme form of the Copenhagen interpretation says the everyday world of certainty comes about when a big thing observes a small thing. The everyday world of big things like canon balls flying through the air and planets circling the Sun is said to obey the laws of 'classical' physics, which are basically laws like Newton's that predict what will happen in any given circumstances with 100 per cent certainty. According to the Copenhagen interpretation, then, the everyday world is created when a classical object observes a quantum object.

But what exactly defines a classical object? Does it have to be a collection of billions upon billions of atoms or just a few hundred? The Copenhagen interpretation is tight-lipped on this, which leaves physicists decidedly uncomfortable.

One thing it is impossible to deny is that quantum theory is successful. In fact, it is arguably the most successful scientific theory ever devised. It has predicted the outcome of all atomic and subatomic experiments to a stunning level of precision. And it has literally made our modern world possible, not only giving us lasers, computers and nuclear reactors but an explanation of why the Sun shines and why the ground beneath our feet is solid. This has left physicists in little doubt that quantum theory is a deep and fundamental theory.

But, surely, if a theory is really deep and fundamental, it should apply to everything in Creation – not only to the world of small things like atoms but also to the world of big things like footballs? Why then is the microscopic world, which dances to the tune of quantum theory, so different from the everyday world of trees and planets and people?*

---

* The success of quantum theory creates another big headache for physicists. Einstein's theory of gravity – the general theory of relativity – has passed every observational test like quantum theory. But it is fundamentally a 'classical' theory, which predicts the future with 100 per cent certainty – for instance, where the Earth will be in its orbit next week. At first sight, this does not seem a problem since there appears to be no overlap between the two theories: quantum theory describes the world of small things such as atoms, while general relativity describes the world of big things like stars and the Universe itself. However, when the Universe was young – in the Big Bang – it was smaller than an atom. In order to understand the origin of the Universe it is therefore necessary to mesh quantum theory and general relativity – to create a 'quantum theory of gravity'. Since one theory is founded on uncertainty and the other on certainty, this is a formidable challenge, to say the least.

We Never Observe a Quantum Thing, We Observe Ourselves

The answer, it turns out, is very subtle. It stems from a crucial observation about the world: we never actually directly see a quantum thing like an atom or a photon of light. What we see instead is its 'effect' on some kind of detector or even the retina of our eye. What we observe, in other words, is not the quantum object itself but the 'record' that the quantum object leaves on a large number of other atoms.

'A detector does not measure an exterior system directly, but rather, through an act of observation, changes the state of its own system,' says the Belgian physicist Sven Aerts. In the case of the eye, for instance, light falls on the cells of the retina and changes them – and it is these changed cells that the brain senses, not the light itself. We think we are directly observing light but we are in fact only observing it indirectly. It is ourselves that we are observing directly. 'All observation is self-observation,' says Aerts.

The American physicist Wojciech Zurek puts it more lyrically. 'What the observer knows is inseparable from what the observer is,' he says.

In the light of this – no pun intended – it is now possible to further sharpen the original question, 'Why is the everyday world so different from the microscopic world of atoms?' We can now say 'Why does the "record" of a quantum object like an atom not show any signs of weird schizophrenic quantum behaviour?'

A Water Droplet That's Only Half There

The American physicist Larry Schulman of Clarkson University likes to answer this question with the example of the 'cloud chamber'. This is an ingenious device that creates a record of the passage of a microscopic particle such as an atom. Crucially, the record is substantial enough to be visible to the eye, so the question of how the record of a quantum event loses its quantumness is no longer academic. It is a nuts-and-bolts question which requires a nuts-and-bolts answer.

Essentially, a cloud chamber is a sealed box containing water vapour with an observation window in its side. Now, water vapour, if cooled, condenses to form visible droplets of water. This is of course what happens when a cloud – a very large volume of water vapour – is forced to rise over a hill or mountain in its path. Because the air is cooler at higher altitude, water vapour condenses into droplets, which fall as rain. However, the formation of a water droplet requires more than simply the cooling of water vapour. There must also be a 'seed' around which the water droplet can grow.* In the atmosphere, the necessary seeds are provided by tiny motes of dust floating in the air. If no such seeds are around, it follows that it is extremely difficult for droplets to form. This observation, it turns out, is the key to the operation of the cloud chamber.

The cloud chamber is filled with water vapour that is ultra-pure. It is so ultra-pure, in fact, that there are no seeds around which droplets can form. This means that, when the water vapour is cooled, water droplets are absolutely desperate to form but they cannot. They are utterly frustrated.

Think of the water vapour in a cloud chamber as like a minefield in which all the mines are on the very brink of exploding. In such circumstances, the tiniest breath of air or vibration of the ground will be enough to set the mines off. Similarly, in the cloud chamber, the tiniest

---

* Actually, this is not strictly true. It is possible for a water droplet to grow spontaneously – that is, without a seed. But this is extremely rare.

thing will be enough to act as the necessary seed for the explosive formation of a water droplet. And 'tiny' is the operative word here. For, remarkably, the seed can be something as small as a single quantum event – for instance, the collision of a high-speed atomic particle with a water 'molecule'.*

Imagine, then, a high-speed atomic or subatomic particle passing through the cloud chamber. It could be a particle fired deliberately into the chamber by an experimenter or perhaps a 'cosmic ray' particle which has come down from space. As it zips through the chamber, the particle will smash into water molecules in its path, kicking free their orbiting electrons. A molecule bereft of some of its electrons is said to be 'ionised' and an ionised molecule, it turns out, is sufficient to act as the necessary seed for the formation of a water droplet. Consequently, a stream of tiny water beads will mark the track of the particle through the cloud chamber. If illuminated properly, the stream will be visible to the naked eye through the window in the cloud chamber.

The cloud chamber was in fact the first device ever built that could reveal the 'tracks' of subatomic particles such as electrons. It was invented by the English physicist Charles Thomson Rees Wilson in 1911. Intrigued by the peculiar clouds that formed above the peak of Ben Nevis in Scotland, he had attempted to create artificial clouds in his laboratory in Cambridge. The fruit of his efforts was the cloud chamber, for which he shared the 1927 Nobel Prize for Physics.

For Wilson it was sufficient that he had miraculously made the atomic realm visible. This was in the days before there was a full appreciation of the bizarre nature of the quantum world so he did not get hung up on the details of precisely how the cloud chamber worked. Nowadays,

---

* Water is made of 'molecules', each of which contains one atom of oxygen bound to two atoms of hydrogen.

however, understanding the way in which the droplets form is the key to understanding how the familiar, everyday world emerges from the bizarre, schizophrenic world that underpins it.

Imagine a water-vapour molecule sitting in the path of the high-speed particle flying through the cloud chamber. The particle can either ionise it or not ionise it. This is obvious and uncontroversial. But we are dealing with an event in the quantum world. And this means that, in addition to there being a quantum wave representing the molecule being ionised and another quantum wave representing the molecule not being ionised, there is a third possibility – a superposition of the two waves. In other words, the water molecule can be simultaneously ionised and un-ionised.

By rights, a water molecule that is simultaneously ionised and un-ionised ought to trigger a curious water droplet – one that hovers half in existence and half out of existence. A water droplet that, like the smile on the face of the Cheshire cat, is only half there.

This is pretty much a real-life enactment of the famous 'Schrödinger's cat' thought experiment. The Austrian physicist Erwin Schrödinger imagined an unfortunate cat sealed in a box with a vial of poison which could be broken by a hammer. The hammer was triggered by the disintegration, or 'decay', of an unstable atom which had a 50 per cent chance of decaying during the experiment. Since quantum theory permits an atom to simultaneously decay and not decay, Schrödinger's question was: 'Before the box is opened and the cat is observed to be either dead or alive, was it simultaneously dead and alive?'

Schrödinger's thought experiment elevated an event in the quantum world into the everyday, large-scale world, forcing physicists to seriously address the crazy implications of their theories. However, it was hardly an experiment that could actually be done (at least, without upsetting legions of cat lovers). Schulman's cloud chamber, though not as striking as Schrödinger's cat, at least has the merit of being a real, nuts-and-bolts, practical device.

So, what of a water droplet that hovers half in existence and half out of existence? It goes without saying that nobody has ever seen such a schizophrenic water droplet while looking through the window in a cloud chamber. But why? Where does the quantum weirdness go?

Well, in order to see weird quantum effects it is fundamental that the individual waves of a quantum superposition overlap each other. If they do not overlap, then they cannot interfere with each other and it is through interference that quantum weirdness occurs.

In the case of the double-slit experiment, for instance, it is never possible actually to observe an atom going through both slits at the same time. Nature, it turns out, is extremely careful to conceal its hand. All we can ever see is the 'consequence' of the atom going through both slits simultaneously – that is, the black-and-white-striped interference pattern on the second screen. In the double-slit experiment, this is the quantum weirdness – and it happens only if the quantum wave representing the atom going through one slit and the quantum wave representing the atom going through the other slit interfere with each other. Which they can do only if they physically mingle – that is, pass through the same region of space, or overlap.

So, does the quantum wave that represents a water droplet that exists and the quantum wave that represents a water droplet that does not exist overlap? This is the key question. If there is indeed an overlap, we should see that a water droplet hovering half in existence and half out of existence can exist. If not, we should never see such a bizarre vision.

The only way to answer the question is to imagine the individual water molecules that make up a water droplet. Say, for the sake of argument, that there is a water-vapour molecule that is painted red –

forgetting for a moment that it is impossible to paint a molecule red! – and that it is ionised by the particle flying through the cloud chamber. Now focus on a nearby water-vapour molecule. Water is denser than water vapour, which means the molecules are closer together than they would be in the vapour. So, if the red molecule is ionised and a water droplet forms, the nearby molecule will be closer to the red molecule than it would be if the red molecule is not ionised and no droplet forms. Say the overlap between its quantum wave in the first and in the second case is 50 per cent. This is still enough for the waves to interfere.

Now consider a second molecule near the red one. Say, once again, the overlap between its wave in the case when the red molecule is ionised and in the case where the red molecule is not ionised is 50 per cent. What this means is that, if we consider the two molecules together, the overlap between their combined waves is $\frac{1}{2} \times \frac{1}{2} = \frac{1}{4}$. If we consider three molecules, it will be $\frac{1}{2} \times \frac{1}{2} \times \frac{1}{2} = \frac{1}{8}$.

Now – and this is crucial – a water droplet big enough to be visible to the naked eye might contain anywhere from millions to billions of molecules. In other words, the overlap between the wave representing the water droplet existing and not existing is $\frac{1}{2} \times \frac{1}{2} \times \frac{1}{2}$ . . . millions or billions of times over. Clearly, this will be a number within a whisker of zero. Essentially, therefore, there is no overlap between the quantum waves. And, if there is no overlap there can be no interference.* It means

---

* Implicit in this remark is something very peculiar and subtle about quantum theory. The quantum wave representing a particle is a mathematical thing which can be imagined spread throughout ordinary three-dimensional space (after all, the square of the wave height at any point in space is the probability of finding the particle at that point). However, if there are two or more particles, *each* particle has, in a sense, its own copy of 'coordinate space'. In other words, the wave representing five particles lives in a space of $5 \times 3 = 15$ dimensions. And there can be interference only if there is a non-zero overlap between the waves in *all* the spaces. This is why the overlap in the example is given by $\frac{1}{2} \times \frac{1}{2} \times \frac{1}{2}$ . . .

that, when we look through the window of a cloud chamber, we will always see the formation of a water droplet or the formation of no water droplet. We will never see a ghostly water droplet, hovering half in existence and half out of existence.

Here then is the super-subtle explanation for why the schizophrenic events of the quantum world never manifest themselves in the everyday world. To do so, they would have to create a visible record, which in practice means affecting millions or billions of molecules. However, as the cloud chamber example illustrates, it is immensely difficult to have a quantum superposition which involves millions or billions of molecules. By the time large numbers of molecules are involved, the quantum waves in the superposition simply will not overlap, the necessary prerequisite for interference and all weird quantum behaviour.

Physicists have a special word for quantum waves which overlap and which are therefore capable of interfering with each other. They say they are 'coherent'. When a quantum object is recorded by a large number of atoms in a cloud chamber or in the environment, this coherence gets lost. The process, not surprisingly, is called 'decoherence' and it is the ultimate reason why we never see chairs or people in two places at once. It is decoherence that creates the familiar world around us from the nonsensical microscopic world that lies beneath.

Decoherence

The example of the cloud chamber no doubt seems esoteric and far removed from everyday experience. However, in essence, the cloud chamber does something very similar to an electronic particle detector or the human eye. For instance, the eye boosts, or 'amplifies', a quantum event – the arrival of a photon of light.* And it does so by letting the event impress itself on a large number of atoms. Inevitably, by the time large numbers of atoms are involved, the crucial overlap between the individual waves of the quantum superposition is lost. Hey presto. Even in the eye, the weird, schizophrenic quantum world is replaced by the well-behaved, everyday classical world.

It is a common belief that quantum theory applies only to the microscopic world of atoms and their constituents and not to the everyday world of big things. But decoherence shows that this is a complete fallacy. In reality, quantum theory is a fundamental theory that applies to all aspects of the Universe. It is simply that its most peculiar aspects are never apparent in the everyday world. They are hidden by the obscuring hand of decoherence.

The instant a microscopic object impresses itself on its environment – be it a cloud chamber droplet or the human eye – the quantum weirdness gets irretrievably lost.† The environment simply fails to record its schizophrenic nature. Think of an individual at a football game whose voice leaves no discernible impression on the roar of the crowd. That is kind of the way it is with quantum events.

It is often said that a quantum superposition is a terribly fragile thing and that, when it interacts with the environment, the environment destroys it. In fact, this is back to front. A quantum superposition is a pretty robust thing. However, when it interacts with its environment, the environment has enormous trouble recording the superposition among its vast number of 'degrees of freedom'. It is actually the environment's ability to record the superposition that is a fragile thing.

---

* Actually, the light-detecting cells in the retina are not up to detecting a single photon. But, remarkably, they are sensitive enough to respond to a mere handful.
† Decoherence is one of the fastest processes known. It is not unusual for it to happen within a 10 million trillionth of a second.

The environment need not be anything as esoteric as a cloud chamber or even as biological as an eye. It can simply be the surrounding Universe. The Moon, for instance, is continually impressing itself on its environment because photons – particles of sunlight – are constantly bouncing off its face and flying away into the night. Sooner or later, they impress themselves on everything in the environment from the Earth to the distant stars to the human eye. And it is in these bodies – large collections of atoms – that any weird quantumness of the Moon is lost.

Einstein asked: 'Is the Moon there when nobody looks?' The answer is yes – because the presence of the Moon is continually being recorded by the Universe itself.

Quantumness is Not the Preserve of Small Things

Because coherence – which leads to quantum weirdness – is lost when an object impresses itself on its surroundings, it follows that quantum weirdness is a property of objects that are isolated from their surroundings. It is the isolation that is the key. Contrary to the popular view, quantum weirdness turns out to have nothing to do with an object being big or small. If it were possible to take a big object and successfully isolate it from its environment, it would continue to behave like a weird quantum thing. In fact, this is exactly what a team led by Zeilinger at the University of Vienna has been trying to show with increasingly large objects. So far, they have succeeded in making a 'buckyball', a football-shaped molecule containing 60 carbon atoms, 'be in two places at once' by going through two slits in a screen simultaneously. For their next trick, they aim to do something similar with a virus, a bundle of protein molecules which straddles the border between life and non-life.

In principle, it is possible to arrange for a human being to walk through two doors simultaneously. What makes it impossible in practice is the severe difficulty of isolating a big thing from its surroundings. This is the reason – a solely practical one – that weird quantum behaviour is almost exclusively associated with small things like atoms. Better experiments will reveal bigger and bigger objects behaving in quantum ways. As Zeilinger puts it: 'The border between classical and quantum phenomena is just a matter of money.'

The difficulty in isolating objects from their surroundings creates a tremendous problem for those who would like to exploit the ability of atoms to do many things at once. If this could be done, it would, for instance, be possible to make a computer that could do many calculations at once. So far, in the quest for such a 'quantum computer', physicists have managed to harness no more than ten atoms – each recording a binary 'bit'.* The huge problem they face is in maintaining coherence within the computer – that is, an overlap between the individual waves of a quantum superposition when that superposition impresses itself on the large number of atoms of the computer. This means keeping the quantum superposition totally isolated from its surroundings, which is extremely hard.

---

* Actually, a quantum computer stores and manipulates quantum bits, or 'qubits'. Whereas a normal bit can represent only a '0' or a '1', a qubit can exist in a superposition of the two states, representing a '0' and a '1' simultaneously. Because strings of qubits can represent a large number of numbers simultaneously, they can be used to do a large number of calculations simultaneously.

Two Steps to Reality!

The Herculean struggle faced by physicists trying to build quantum computers is a perfect illustration of how difficult it is for quantum superpositions to survive once large numbers of atoms become involved. And this explains why schizophrenic superpositions can be a central feature of the small-scale world of atoms yet can never be seen in the large-scale everyday world. But does it really explain why the large-scale world of trees and people and planets and stars is so very different from the microscopic world of atoms and their constituents – where the everyday world comes from? The answer is – not completely.

All that decoherence really explains is why, in the world around us, we do not see weird quantum superpositions – for instance, a water droplet that both exists and does not exist at the same time. Being able to rule out this extraordinary possibility is undoubtedly important. But there remain two other, ordinary possibilities. Namely, a droplet that exists and a droplet that does not exist. What determines which of these possibilities actually occurs?

Here we come to a place where science is still all at sea. There is no universally agreed explanation for what rules out one possibility contained in the quantum wave rather than the other possibility. Arguably the most extraordinary explanation, however, was proposed by Princeton graduate student Hugh Everett in 1957. According to his 'Many Worlds' idea, mentioned earlier, all the possibilities encapsulated in the quantum wave in fact occur. At first sight, this seems to fly in the face of our experience. After all, when we look at a particular location in a cloud chamber, we either see a droplet or we see no droplet. If, for instance, we see a droplet, then surely the possibility of a droplet not forming has ceased to exist? No, claimed Everett.

In the Many Worlds, reality, like a forked road, splits into two. In one reality, there is a version of you who sees a droplet form and in the other reality a version of you that sees no droplet form. According to Everett, the quantum wave is not merely an abstract mathematical convenience, useful for calculating things like the probability of something happening. It is completely and utterly real and every possibility encapsulated in it in fact occurs in some reality. Of course, we experience one reality only, and are utterly unaware of all the other realities. And it goes without saying that no one has the slightest idea where all the alternative realities are!

The Many Worlds may sounds like science fiction. However, it may shed light on the operation of quantum computers. Remember, quantum computers exploit the ability of particles such as atoms to be in many places at once to do many calculations at once. The field is still in its infancy and the capabilities of quantum computers are puny. However, there is no reason to believe that, within the next twenty or thirty years or so, more powerful quantum computers will not be built. Such computers could solve in seconds certain problems that would take a conventional computer longer than the age of the Universe.

And herein lies one of the central puzzles posed by quantum computers. Extrapolating into the far future, it is easy to imagine a quantum computer so powerful that it can carry out more calculations at any one time than there are particles in the Universe. An interesting question will then arise. Where will such calculations actually be carried out? After all, if a quantum computer is doing more calculations at any instant than there are particles in the Universe, the Universe simply does not have the physical resources at its disposal to do what the computer is doing.

The Many Worlds provides a natural, but mind-boggling, answer to the conundrum. A quantum computer is never short of the resources it needs to carry out its calculations because it does not have to rely on a single universe. Different parts of its calculations are performed in different realities. Bizarre as it seems, quantum computers may achieve what they achieve by exploiting huge numbers of versions of themselves in other neighbouring realities!

What is unique about the Many Worlds is its view of the quantum wave as a totally concrete thing, with all of the possibilities contained within it equally real. This is not, however, the view of the majority of physicists. They believe that the quantum wave is merely an abstract mathematical thing, useful only in calculating what is likely to happen in any given real-world situation. Consequently, they do not worry about the quantum possibilities that are not actualised. They are not anything anyway, they say. Quantum theory provides physicists with the pro-bability of a particular outcome – the probability of seeing a droplet or no droplet. When one possibility is actualised, there is no need to puzzle over where the other possibility has gone to – as Everett did when he conceived the idea of alternative realities. The other possibility was never anything more than a mathematical possibility anyway.

Whatever your view, it appears that two steps are involved in creating the everyday world from the quantum world beneath. There is the process that stops the individual waves of a quantum superposition from overlapping and interfering. This suppresses quantum weirdness and prevents the existence of, among other things, cats that are both alive and dead. And there is the process which selects one component of the quantum wave. This selects one possibility from all the possibilities present in the quantum wave.

Why Do Big Things Not Get Blurry with Time?

So much for why we do not see big things like people and planets in two or more places at once, but why do we not see big things like trees or footballs becoming more smeared out as time passes? After all, the quantum wave associated with them ought inevitably to spread out with time.

Take that football being struck from the penalty spot again. If it smears out with time, this is equivalent to the football taking several paths through the air simultaneously. Say, for simplicity it goes along just two paths, each with a 50 per cent probability.

Now – and this is the key – the football taking two paths through the air has consequences which are visible to a spectator if and only if the waves representing the two possibilities produce a record on their eyeball with sufficient overlap that there can be interference. Interference, after all, is the prerequisite of seeing quantum weirdness. In practice, however, this would require there being coherence between the waves representing billions upon billions of atoms in the spectator's eyeball. This is impossible. And so the spectator sees the football fly along one and only one trajectory with 100 per cent certainty, obeying Newton's laws of motion and not some averaged form of the Schrödinger equation. Thus is the everyday world saved, once again, from the madness of the quantum world.

5
No Time Like the Present

Why do we experience a 'now'? Because all other ways of experiencing reality would lead to us starving

> He has departed from this strange world a little ahead of me. That means nothing. People like us, who believe in physics, know that the distinction between past, present, and future is only a stubbornly persistent illusion.
>
> Albert Einstein

> I think we can all agree, the past is over.
>
> George W. Bush, May 2000

Deep in the Amazon rainforest, a tree frog sits on a log watching a fly. A genetic fluke has furnished the frog with a brain that perpetually perceives its surroundings as they were one second before. When the fly comes within range, the frog lunges. But, with its out-of-date data, it misses. Weakened by hunger, it falls off the log and dies.

It is a heartbreaking story. However, if you think it is entirely fanciful, think again. A prominent American physicist believes it goes to the very heart of why we perceive the world the way we do. According to James Hartle of the University of California at Santa Barbara, there are a multitude of possible ways we could experience reality – even with a one-second time lag like the unfortunate Amazonian tree frog. 'However, evolution by natural selection has ensured that people and frogs experience the world in the most effective way for their survival,' he says. 'A frog that uses the most recent data to calculate the trajectory of a fly eats; one that doesn't, starves.'

Probably, this seems little more than common sense. However, Hartle, a scientific collaborator of Stephen Hawking's, believes it addresses a major scientific puzzle: why we experience a past, present and future at all. 'The peculiar thing is that none of these concepts is uniquely defined in our most fundamental description of physical reality,' he says.

Einstein's View of Reality

That fundamental description of reality is the special theory of relativity. According to the theory, proposed by Albert Einstein in 1905, space and time are not separate things, as they appear to us to be. Instead, they are blended together into a four-dimensional amalgam called space-time.[*]

In Einstein's revolutionary picture, a person's 'time' is defined by a particular direction in space-time whereas their 'space' is defined by the three perpendicular directions in space-time. The common-sense-defying thing is that there is no unique way of carving up space-time into space and time. Instead, there is an infinite number of possible ways, each of which corresponds to a different notion of time and space.

What determines the way in which someone's space-time is divided up is the direction they are travelling in space-time. If you cannot quite visualise 'direction in space-time', do not worry. It is impossible. As lowly

---

[*]   Space-time is four-dimensional because it blends together one dimension of time and three of space. The time dimension extends of course in the direction past–future while the three dimensions of space are in the directions east–west, north–south and up–down.

three-dimensional beings, we can directly perceive neither four-dimensional space-time nor 'directions' in four-dimensional space-time. Instead, two people travelling in different directions in space-time appear as two people travelling at different speeds.

How exactly will their notions of time and space differ? Well, the first thing to say is that the only meaningful way of measuring intervals of time is with a clock and intervals of space with a ruler. It follows that differences in the time and space of the two people travelling at different speeds will manifest themselves in the properties of their clocks and rulers. As Einstein was first to realise, 'moving' clocks run slow and 'moving' rulers shrink in the direction of motion. Consequently, each person will see the clock of the other running slow and the ruler of the other shrink!

At first sight, it may seem bizarre that each observer sees the same kind of thing when they look at the other. But that is the magic – and also the mind-bending peculiarity – of special relativity. One of the foundation stones of the theory is the recognition that it is impossible to tell who is moving and who is stationary. Imagine sitting on a train and watching another train go by. How can you tell whether it is your train or the other train that is moving? You cannot. As Einstein realised, relative motion alone has meaning. It is possible only to say that you are moving at such-and-such speed relative to someone else. And, since two observers moving with respect to each other each see the other moving with the same relative speed, each see the other's clock slow and ruler shrink (which is just the same as saying that all ways of carving up space-time into time and space are equally valid).

The fact that the clocks of two observers moving with respect to each other run at different rates means that the two observers will fail to agree on a number of quite basic matters. For instance, they will not agree on whether two events – say, the firing of two widely separated cannons –

occur at the same time. Though one observer may see the cannons go off simultaneously, the other will see them go off one after the other. And this is not all. The two observers will also not agree on 'what is happening now', 'what happened ten minutes ago', and so on. In fact, if the cannons are so far apart that a light ray cannot span the distance separating them in the time between the explosions, the two observers may not even agree on whether one cannon was fired before the other! While one observer sees the one cannon fire first, the other will see it fire second – 'And, if the two observers cannot agree on such basic things, they will not be able to agree on what is past, present and future,' says Hartle.

So why, contrary to special relativity, do we have such a strong feeling that the past, present and future are uniquely defined? The answer is that noticeable differences in the rate of clocks occur only if observers are moving relative to each other at speeds approaching that of light itself.* However, everything we do on Earth involves speeds much slower than that of light – even a passenger airplane flying through the air travels more than a million times slower than a light beam in empty space. 'In everyday situations, it turns out the differences in our notions of time are too small to notice,' says Hartle. 'This is the ultimate reason we can all agree on a past, present and future.'

The fact that we are condemned to live our lives in the cosmic slow lane may explain why we experience a common past, present and future. However, it does not explain why the present has such a special

---

* The second foundation stone of special relativity – in addition to the idea that only relative motion has meaning – is that the speed of light is always the same. In fact, the reason that time and space distort when relative speeds approach that of light is specifically so that all observers, no matter what their speed, measure exactly the same speed for a light beam. Space and time are but shifting sand; the speed of light turns out to be the bedrock on which the Universe is built.

immediacy to us. Why do we focus our attention on the most recent information we have gathered from our surroundings – what is happening right 'now'? Why do we not focus on information gathered one second ago like the unfortunate tree frog? Or ten seconds ago? Or half an hour ago, for that matter?

This is a tough question to address. The human brain, after all, interacts with its surroundings in an extremely complex way. However, in the time-honoured manner of theoretical physicists, Hartle attempts to answer the question from the point of view of a more rudimentary entity than a human: a robot which 'experiences' and 'reacts' to reality in the simplest way imaginable.

Robots that Mimic Humans

The robot Hartle envisages is an 'information gathering and utilising system', or IGUS. This is an abstract entity first imagined by Nobel Prize-winning physicist Murray Gell-Mann of the California Institute of Technology in Pasadena. An IGUS has an 'input register' in which it records information from its environment. This could, for example, be an image of the surroundings. The storage capacity of the input register is limited, however. In order to make room for new information continually flooding in from the environment, the information is passed after a while to a 'memory register'.

The robot might have many memory registers along which it continually shuffles past information. But it does not have infinite resources at its disposal. So, eventually, the registers come to an end and the information is unceremoniously dumped. However, simply dumping the information without first extracting anything useful from it is wasteful.

Before the information is thrown away, therefore, it is passed on to other parts of the robot. And this is where the important stuff goes on.

Hartle's robot, in addition to having storage registers, has a 'schema' and a 'computer'. The schema is a simplified model of its environment plus a set of rules, culled from the robot's past experience, which tell it how to behave in particular circumstances. The computer is, well, just a computer. However, it carries out two quite distinct types of computation. First, there are 'unconscious' computations, which update and improve the model of the environment stored in the schema. Then there are 'conscious' computations. These determine how the robot should respond to information pouring in from its surroundings, based on the rules stored in its schema.

The robot probably seems a bit on the complicated side. However, Hartle claims that a robot like this, carrying out unconscious and conscious computing, mediated by a schema, can imitate some of the key features of human perception. And, crucially, it has the advantage of being fantastically simpler to deal with than a human being.

For the robot to mimic a human being, says Hartle, it is necessary only for it to carry out its 'conscious' computation exclusively on the contents of its input register and its 'unconscious' computation on its memory registers. 'This distinction is very important,' says Hartle. 'It ensures that – just like a person – the robot consciously "experiences" the present but only "remembers" the past.'

Hartle maintains that a robotic IGUS set up this way also experiences the world in other ways that are like a person's. For instance, while the past in the registers is 'remembered', the future is 'predicted'. The future is the result of computation – for instance, the computation that a car will hit you in a second's time unless you take avoiding action. The past and future are therefore qualitatively different from each other, just as they are for us. 'Even God cannot change the past,' said the Greek dramatist Agathon. Furthermore, because such a robot focuses its attention on the input register – that is, on the information most recently acquired from its surroundings – the 'now' has a very special immediacy.

This may all seem terribly abstract. However, in practice, a human-mimic IGUS works quite simply. Say, for instance, an image of a cheeseburger appears in the first register. According to Hartle: 'The computer consults the schema, which has abstracted rules from a previous experience – from previous visits to burger restaurants – and realises "Hey, I like cheeseburgers." The robot therefore decides to buy a cheeseburger. Or perhaps the schema contains information on the fat content of burgers, which overrides the liking of burgers, so the computer decides not to buy a cheeseburger after all!'

Hartle claims this human-mimic IGUS sheds light on the long-standing puzzle of why we seem to experience a 'flow of time', even though this cannot be so. 'Something which flows must, by definition, change with time,' says Hartle. 'But how can time change with time? It's a logical impossibility.'

The impossibility of a flow of time is in fact explicit in special relativity. According to Einstein, space-time is like a 'map' in which all the 'events' in the history of the Universe – from the birth of everything in the Big Bang way into the far future – are laid out, exactly as if they are pre-ordained. Nothing at all flows.

In the human-mimic IGUS, however, images pass from memory register to memory register until finally they are dumped, or erased – the robot's equivalent of 'forgetting'. It is this, says Hartle, which represents the 'flow of time'. It is a flow of information. 'Something like this flow of information from register to register must be happening in our brains,' he says. 'Ultimately, this is what we interpret as the flow of time.' As Austin Dobson wrote in 'The Paradox of Time':

Time goes, you say? Ah no!
Alas, Time stays, we go.

All this IGUS stuff may seem a bit complicated in order to simply conclude that we are interpreting the flow of information from neurone to neurone in our brains as the flow of time. Fortunately, there is a pay-off. The great beauty of Gell-Mann's IGUS is that it is extremely flexible. By wiring it up in different ways, it is possible to change the flow of information between its components. This enables a robotic IGUS to experience and react to reality in different ways – ways which can be very different from a human being.

This capability enables Hartle to ask a key question: 'Are there other ways that creatures could organise their experience – ways that are different from ours but still consistent with the basic laws of physics?'

Other Ways of Perceiving Reality

To try and answer the question, Hartle imagines several different types of robot, each of which experiences reality in a different way. The first robot focuses on not one but two times, ten seconds apart. In other words, it has two 'presents'. A second kind, much like the Amazonian tree frog, is always behind, seeing the world as it was a few seconds ago. And a third type of robot has no 'schema'. It therefore has no unconscious thought and no simplified model of its environment. Its next move must be computed from contents of all its registers – it has no way of focusing only on relevant information. 'The question is: would any of these creatures be viable?' asks Hartle.

According to Hartle, the two-time robot would waste valuable computing resources by consciously focusing on unessential information in the past. The always-behind robot would starve to death just like the tree frog. And the no-schema robot would squander even more precious computational resources than its two-time cousin. 'Should creatures ever arise with any of these variant ways of organising their experience, they would be weeded out by natural selection,' says Hartle. 'Outcompeted by creatures who experience reality much like us, they would pretty quickly become extinct.'

Remarkably, Hartle concludes: 'The way we experience time is determined as much by the laws of biology as the laws of physics.' This turns out to have implications for any extraterrestrials we may one day meet. Hartle believes he can say categorically that they will experience the world in exactly the same way as us, sharing the concepts of past, present and future, and the idea of a flow of time.

There is, however, a way that creatures could have arisen that organise their experience differently from us, according to Hartle – if the laws of physics were different. 'Say, the force on a massive body depended not on the force acting on it and the position of the body, as it does in the world we live in, but on its position now and 10 seconds ago,' says Hartle. 'In such circumstances, natural selection would clearly favour the evolution of a split-time creature with two presents.'

Evolution by natural selection may have made us the way we are and given us our concept of time. Nevertheless, we have the technology to build robots any way we like. This raises a bizarre possibility, which Hartle considers. 'Might it be possible to build a robot which organises its experience back-to-front?' he says. 'A robot, in other words, that remembers the future and predicts the past?'

According to Hartle, it is not difficult to imagine a robot that is in a position to remember the future. Images would simply have to flow backwards along its registers – say, from right to left rather than from left to right. This would involve 'unerasing' information that had been dumped, or erased. Unerasing information is the equivalent of re-assembling a piece of paper that has gone through a shredder. 'It is difficult and it costs energy but it is not impossible,' says Hartle. 'It is perfectly possible to conceive of building a robot with a reverse-time sense.'

The really difficult problem, however, turns out to be giving the robot a future environment to remember! This is because of the difficulty in reversing the so-called arrow of time.

Large collections of atoms such as people and medieval castles grow old and crumble as time passes. These transformations are associated with a change from order to disorder. Physicists associate this increase in disorder, or 'entropy', with the so-called thermodynamic arrow of time. The direction of the arrow – from past to future – is associated with the direction in which things become disordered, in which people grow old and medieval castles crumble to dust. Consequently, if a robot is to remember the future rather than the past, the arrow of time must be reversed in its surroundings so that people grow young and medieval castles uncrumble.

It goes without saying that this is extremely difficult. 'The past is a foreign country. They do things differently there,' wrote L. P. Hartley in *The Go-Between*. According to Hartle, reversing the arrow of time would mean reversing the velocity of every particle of matter in the robot's neighbourhood. Even doing this to the air molecules flying about a room would be unimaginably harder than unshredding a shredded piece of paper. 'Also, all matter interacts with light, so simply to reverse the arrow of time for a day, it would be necessary to deal with every molecule and particle of light within the distance light travels in a day – a radius of 26 billion kilometres,' says Hartle. 'Perhaps a super-advanced civilisation might find it amusing to do. But it's way beyond our capabilities.'

Experiencing the Universe's More General Reality

As noted earlier, we experience reality in the ultra-slow lane where the full effects of special relativity are not felt and it is possible to agree on a 'common' past, present and future. But what about the more general situation when the full effects of relativity are important and peculiar things happen to time? Well, it turns out that the satellites that make up the Global Positioning System can be considered as an IGUS set up to experience time in a peculiar way.

People using GPS rely on receiving signals from clocks on the different satellites. The receiver determines its position on Earth by comparing the different amounts of time the various signals take to reach it. The system depends on the clocks on the satellites all ticking at the same rate. However, they do not. As special relativity predicts, time slows down at high speed. In fact, it is worse than this. Einstein's general theory of relativity, which incorporates special relativity, predicts that time also slows for bodies experiencing strong gravity. Since gravity is stronger closer to the Earth time slows down for a GPS satellite when it swoops close to the surface.

The 'relativistic' slowing of time experienced by the satellites is tiny – one clock may slow by only a few nanoseconds per second relative to another. However, measuring time delays to nanoseconds turns out to be crucial if a receiver is to determine its position on the surface of the Earth to an accuracy of a few metres.

So much for the details of GPS. The point, as far as the experiencing of time is concerned, is that all the satellite-borne clocks are running at slightly different rates. And this, in turn, means that they each have different perceptions of 'now'. In fact, for the system to operate, the GPS engineers had to design a particular meaning of 'now', which is different from the everyday meaning. 'In other words, they tinkered with the system's perception of time,' says Hartle.

But the familiar concepts of past, present and future not only break down at speeds comparable to light – for instance for the GPS satellites – they break down in other circumstances too. Imagine that it is possible to perceive things which happen in ultra-small intervals of time – say, in less than a billionth of a second. This would radically affect your perception of a friend you are talking with. Why? Because light takes about a billionth of a second to bring a picture of them to you.

Such an interval is ten million times shorter than any event that can be perceived by the human brain, so it is unnoticeable in normal circumstances. However, if you have ultra-fast perception, you do not hear what your friend is saying 'now' but what they said in your past. The two of you do not share a 'now'. And people standing farther away from you appear even farther in the past. No longer is there a common present.

The other situation where the concept of past, present and future breaks down is when observers are separated by a large distance compared with the distance light can travel in the time it takes any events of interest to occur. This would be the case, for instance, for a galaxy-spanning empire, which, like a human being and like the GPS system, can be considered an IGUS. 'There would be no point in defining a "now" on a planet at the centre of the Galaxy when light takes 60,000 years to take knowledge of it to a planet on the periphery,' says Hartle. 'Clearly, such a civilisation would need to organise its time differently from us.'

Time is often considered a slippery, elusive thing. 'What, then, is time?' St Augustine wrote in the fifth century. 'If no one asks me, I know what it is. If I wish to explain what it is to him who asks me, I do not know.' Hartle, however, disagrees that time is elusive at all. 'All special relativity teaches us is that the idea there is "one" time is an illusion,' he says. 'Rather there are many.' And, just because special relativity does not say anything about the past, present and future, does not mean they are illusions – they are real, strongly held properties, of an IGUS like a human being. 'Newtonian physics doesn't tell us how many continents there are on the Earth', says Hartle. 'But the continents are real none-theless.'

Perhaps the final word should go to the nineteenth-century British essayist Charles Lamb: 'Nothing troubles me more than time and space,' he wrote. 'And yet nothing troubles me less, as I never think about them!'

6

God's Number

Where can we find the secret of the Universe? In a single number!

Computers are useless. They can only give you answers.
Pablo Picasso

There is hardly any paradox without utility.
Gottfried Leibniz, letter, September 1695

Some time ago a group of hyper-intelligent pan-dimensional beings decided to answer the great question of Life, The Universe and Everything. To this end they built an incredibly powerful computer, Deep Thought. After the great computer program had run for a few million years, the answer was announced. And the answer was . . .

.0000001000000100001000001000011101110011001001111000
10010011100 . . .

Come again? Surely, it was forty-two? Well, in Douglas Adams's novel *The Hitchhiker's Guide to the Galaxy* it certainly was. But, in the real world, rather than the world of Arthur Dent, Zaphod Beeblebrox and

Ford Prefect, the answer to the question of Life, The Universe and Everything is very definitely . . .

.0000001000000100001000001000011101110011001001111000
10010011100 . . .

The number is called Omega and, remarkably, if you knew its first few thousand digits, you would know the answers to more mathematical questions than can ever be posed. What is more, the very existence of Omega is a demonstration that most mathematics cannot be discovered simply by applying logic and reasoning. The fact that mathematicians have little difficulty in discovering new mathematics may therefore mean that they are doing something – employing 'intuition' perhaps – that no computer can do. It is tantalising evidence that the human brain is more than a jelly-and-water version of the PC sitting on your desktop.

Omega (Ω) actually crops up in a field of mathematics invented by an Argentinian-American called Gregory Chaitin. Algorithmic Information Theory attempts to define 'complexity'.[*] This is a very difficult concept to pin down precisely yet a precise definition is extremely important in many fields. How else, for instance, can a biologist studying evolution objectively say that a human is more complex than a chimpanzee or even a jellyfish?

Chaitin invented AIT when he was fifteen, the same age Wolfram was when he began publishing papers in physics journals. Chaitin's principal concern at the time was with numbers. But, in fact, AIT applies to much more. After all, as we all know today, information describing everything – from words to pictures to music – can ultimately be expressed in the form of numbers. We are living in a 'digital' world.

[*] Many of the ideas of AIT were invented independently by the Russian Andrei Kolmogorov.

Chaitin's key idea was that the complexity of a number can be measured by the length of the shortest computer program that can generate it. Take, for instance, a number which goes on for ever such as 919191 . . . Although it contains an extremely large number of digits – it goes on for ever after all – it can be generated by a very short program:

Step 1: Write '91'
Step 2: Repeat step 1

According to Chaitin's measure, therefore, the number 919191 . . . is not very complex at all. The information it contains can be reduced, or 'compressed', into a much more concise form than the number itself – specifically, the two-line program above.

Actually, Chaitin is a bit more precise about what he means by the 'shortest computer program'. He is a mathematician, after all. He means the 'shortest computer program encoded in binary that can generate a particular number, itself expressed in binary'.

Binary means a string of 0s and 1s and is pretty much synonymous with the word digital.* All computer programs – including Microsoft Windows – are ultimately encoded in binary. So it is not hard to imagine a computer program with strings of 0s and 1s representing both numbers and commands such as 'Repeat step 1'. It is the length of just such a program that Chaitin equates with the complexity of a number.

* Binary was invented by the seventeenth-century mathematician Gottfried Leibniz. It is a way of representing numbers as a string of 0s and 1s. In everyday life, we use decimal, or base 10. The right-hand digit represents the 1s, the next digit the 10s, the next the 10×10s, and so on. For instance, 9217 means $7 + 1 \times 10 + 2 \times (10 \times 10) + 9 \times (10 \times 10 \times 10)$. In binary, or base 2, the right-hand digit represents the 1s, the next digit the 2s, the next the 2×2s, and so on. So, for instance, 1101 means $1 + 0 \times 2 + 1 \times (2 \times 2) + 1 \times (2 \times 2 \times 2)$, which in decimal is 13.

According to AIT, if there are two numbers and generating the first requires a program the length of 37 binary digits, or bits, while generating the second requires one 25 bits long, the first number is the more complex.

Pattern is the key. If the digits of a number have some kind of pattern – like the pattern of 919191 . . . – the pattern can be used as a shortcut to generate the number. Consequently, the binary program necessary to generate the number is relatively short – it has fewer bits than the number itself. Such a number is said to contain reducible information because it can be reduced, or compressed, into a more compact form – the form of the computer program.

Most numbers have no discernible pattern, however. Unlike 919191 . . . , their digits are entirely unrelated to each other. The only way for a computer program to generate such a number is to write it out in full. This is no compression at all. The program is as long as the number itself. Such a number is said to contain irreducible information because it cannot be squeezed into a more compact form. This is where Omega, 'the jewel in the crown of AIT', comes in. It takes the idea of irreducible information to its insane, logical extreme.

Omega, which was first defined by Chaitin in the 1970s, is an infinitely long number whose digits are without the slightest trace of pattern. Consequently, there is no way to generate its first 10 digits with a program less than 10 digits long; no way to create its first 511 digits with a program shorter than 511 digits in length; and so on. Omega's never-ending sequence of 0s and 1s can be generated only with an infinitely long computer program. There is no shortcut, no way to compress it into a more compact form. It is the ultimate in irreducible

---

★ Actually, there is not one Omega but a whole class of Omegas. This is because Omega depends on the particular type of computer language used to generate a number. It would not be the same, for instance, in two languages that used a different string of 0s and 1s to code for a task like 'Repeat step 1'.

information.* Chaitin calls it a 'very dangerous number'. 'On the one hand, it has a simple, straightforward mathematical definition,' he says. 'On the other hand, its actual numerical value is maximally unknowable.'

People often think of pi (π) – the ratio of the circumference to the diameter of a circle – as complex. After all, its digits – 3.1415926 . . . – go on for ever and do not appear to repeat. However, it turns out that pi can be generated by a relatively simple computer program and so, by Chaitin's measure, is not very complex at all. 'By comparison, Omega is infinitely more complex,' says Chaitin.

Omega is like an infinite series of coin tosses – with the heads equivalent to 0s and the tails to 1s. The outcome of each toss in such a series is entirely unrelated to that of the previous toss. The only way to discover the sequence of heads and tails in an infinite series of coin tosses is to toss a coin an infinite number of times. There is no shortcut. 'And this is exactly the way it is with the digits of Omega,' says Chaitin.

But Omega, it turns out, is more – much more – than simply an infinitely random, infinitely complex, infinitely incompressible number. Unexpectedly, it turns out to have a deep connection with the ultimate limitations of computers – what they can and cannot compute.

Uncomputability

The question of what computers can and cannot do was an obsession of the English mathematician Alan Turing. During the Second World War, Turing was stationed at Britain's top-secret code-breaking establishment at Bletchley Park in Buckinghamshire. There he helped break the fiendishly complex 'Enigma' and 'Fish' codes with which the Nazis encoded their most secret radio transmissions. The intelligence gathered enabled Winston Churchill and the Allies to anticipate German actions, saving countless lives by shortening the war, some say by up to two years.

Turing's code-breaking success relied on 'Colossus', the world's first programmable electronic computer, more than ten of which were in operation at Bletchley Park and Cheltenham by the end of the war. But his enduring fame rests on work he carried out earlier, in the 1930s, on a far more theoretical type of computer – one he invented with the specific purpose of figuring out the limits of computers.

A 'Turing machine' is simply a box. A one-dimensional tape with a series of 0s and 1s inscribed on it is fed into the box and the same tape emerges from the box with a different series of 0s and 1s on it. The 'input' is transformed into the 'output' by a read/write head in the box. As the tape passes the head, one digit at a time, the head either leaves the digit unchanged or erases it, replacing a 0 by a 1, vice versa. Exactly what the head does to each digit is determined by the 'internal state' of the box at the time – what, in today's jargon, we would call a computer program.

With its input and output written in binary on a one-dimensional tape, a Turing machine is a wildly impractical device. Practicality, however, was not the point. The point was that, with the Turing machine, Turing had invented – on paper, at least – a machine that could simulate the operation of absolutely any other machine.[*]

Nowadays, a machine that can simulate any other machine – a 'universal machine' – is not considered remarkable at all. Such devices – capable of carrying out not one specialised task but any conceivable task – are ubiquitous features of the world. They are called computers. In the 1930s, however, the universal Turing machine appeared to be straight from the pages of science fiction. The only way computing machines of

[*] And it could do this using only seven basic commands: (i) Read tape, (ii) Move tape left, (iii) Move tape right, (iv) Write 0 on tape, (v) Write 1 on tape, (vi) Jump to another command, and (vii) Halt.

the day could carry out different tasks was if they were painstakingly rewired. Turing's genius was to see that this was unnecessary. With a universal machine – a general-purpose computer – it was possible to simulate any other machine simply by giving it a description of the other machine plus the computer program for the other machine. There was no need to change the wiring – the hardware – only the software.

Turing imagined the software for his universal Turing machine inscribed as a long series of 0s and 1s on a cumbersome one-dimensional tape. Fortunately, today's computers are a bit more sophisticated than Turing's vision and their software comes in a considerably more user-friendly form.

In the universal Turing machine, however, Alan Turing can in fact be said to have invented the modern general-purpose computer, a machine whose unprecedented flexibility is guaranteed by infinitely re-writable software. His genius was to recognise that, in the final analysis, all a computer really is is a device for shuffling symbols. One sequence of symbols is fed in. And another sequence of symbols is spat out, which depends on its computer program. This is all any computer does. Take, for instance, the computer that flies a plane. It is fed sequences of symbols which tell it the position of the plane, its speed, the engine temperature, and so on. The program then acts on this information, telling the computer what sequences of symbols it should spit out in order to control engine revs, rudder direction and so on.

A universal Turing machine similarly was a symbol shuffler. But the key thing for Turing was that it could simulate the operation of any conceivable machine. This meant that it could compute anything that was computable and, recall, he was interested in what could and could not be computed. All he had to do, therefore, was find a task that would flummox a universal Turing machine. Remarkably, he stumbled on one immediately. Even more remarkably, the task was fantastically simple.

The impossible task was to take a computer program – any computer program – and determine whether or not it ever stops.

What exactly does this mean? Well, people who write computer programs for a living know that such programs sometimes get stuck in endless loops, going round and round the same set of instructions like a demented hamster in a wheel. The task Turing set his universal Turing machine was to take a program and compute whether or not this will happen – spitting out, say, a 0 if the answer is yes, and a 1 if it is no.

At first sight, this 'halting problem' seems ridiculously simple to solve. The easiest way to check whether a particular program halts or not is simply to run it on a computer and see. This is certainly feasible if the program comes to a halt after a minute or an hour or even after a year. But what if it halts only after 1,000 years or a trillion trillion years? Nobody can wait that long. Now, perhaps, the deceptive trickiness of the halting problem is apparent. It is about taking a program and computing in advance of actually running it whether it will 'eventually' stop.

And the remarkable thing that Turing discovered was that this apparently simple task is impossible. Though it is easy to state, no conceivable computer, no matter how powerful, can ever compute it.*

---

* The halting problem is uncomputable because, if there was a program that could compute it – one that could take another program and spit out, say, a 0 if it never halts and a 1 if it eventually halts – this 'halting program' could be used to do something impossible: construct a program which stops if it doesn't stop and doesn't stop if it stops!

How? Such a program would have to incorporate the halting program as sub-program, or subroutine, and apply it to itself. This sounds tricky but actually is not. Just engineer the program to output itself – a string of 0s and 1s identical to the binary code of the program – and then get the halting program to check whether the output halts or not. If it does halt, the program has instructions not to halt; and if it does not halt, the program has instructions to halt.

What has been concocted is an impossibility, a contradiction – all made possible by the existence of the halting program. For the sanity of the Universe, therefore, the halting program cannot exist.

The halting problem would appear to have nothing whatsoever to do with Omega (though, like the halting problem, Omega is simple to define but impossible to compute). The two are, however, very deeply connected. Omega, it turns out, is more than irreducible information, naked randomness. Omega is the 'probability' that a randomly chosen computer program – one picked blindly from all possible computer programs – will eventually halt.

Probabilities are conventionally written as fractions between 0 and 1, with a probability of 0.5 corresponding to a 50 per cent chance of something happening, a probability of 0.99 to 99 per cent and so on. Since Omega is defined as a probability, there is in fact a decimal point before its first digit, as can be seen in the Omega written out (in part) at the beginning of this chapter.

But what does it mean that Omega is the probability that a randomly chosen computer program halts? Well, think of generating a string of 0s and 1s by repeatedly tossing a coin. Such a string can encode the instructions of a computer program just as easily as it can encode the bars of a piece of music or the picture elements of a family photo. Well, Omega is the probability that a computer program generated in a random manner like this will eventually halt.

Put it another way. Omega is what you get when you take all possible computer programs that can exist, one at a time. You see whether or not each halts, giving a 0 for one possibility and a 1 for the other. Because there is an infinite number of possible programs, there will be an infinite number of 0s and 1s. Well, you take the average of all the 0s and 1s. And that is Omega.

Put it another way. Omega is the concentrated distillation of all conceivable halting problems. It contains the answer not just to one halting problem but to an infinite number!

Of course, individual cases of the halting problem are uncomputable.

This was Turing's big discovery, after all. Consequently, Omega too is uncomputable. This is in fact not very surprising. Recall that it takes an infinitely long computer program to generate Omega, which is hardly a practical proposition.

Omega is maximally uncomputable, maximally unknowable. 'Technically, this is because the first n bits would in theory enable us to answer Turing's infamous halting problem for all programs up to n bits in size – and this is impossible,' says Chaitin. 'However, crudely speaking, the reason Omega is unknowable is that it's the probability of something happening – a computer program halting – which itself is unknowable!'

The deep and unexpected connection between Omega and all conceivable halting problems has an astonishing consequence. 'It comes about because of the remarkable fact that most of the interesting problems in mathematics can be written as halting problems,' says Cristian Calude of the University of Auckland.

Take, for example, the problem of finding a whole number that is not the sum of three square numbers. The number 6, for instance, can be written as $1^2 + 2^2 + 1^2$ and so is the sum of three square numbers. The first number that is not a sum of three squares is in fact 7.

A brute-force program to find numbers that are not the sum of three squares would simply step through the whole numbers, one at a time, stopping when it finds a number that cannot be written as the sum of three squares. Or, if all numbers can be written as the sum of three squares, it will keep going for ever. 'Does this ring any bells?' says Calude. 'It's a halting problem!'

The amazing thing is that a host of other mathematical questions can also be re-cast as halting problems – if a particular program halts, the answer to the question is yes; if it doesn't halt, it is no. The consequence of this fact is scarcely believable. 'The first few thousand digits of Omega contain the answers to more mathematical questions than could be written down in the entire universe!' says Charles Bennett of IBM in New York.

An example of such a question is whether Fermat's Last Theorem is correct. This was inserted in the margin of a book by the French mathematician Pierre de Fermat in 1642 and finally proved three centuries later by Andrew Wiles in 1995. It asserts that there are no positive whole numbers – call them x, y and z – for which $x^n + y^n = z^n$, if n is bigger than 2. There are, for instance, no whole numbers x, y and z for which $x^3 + y^3 = z^3$ or $x^{99} + y^{99} = z^{99}$. Another example of a mathematical question is whether Goldbach's Conjecture is correct. This states that every even number greater than 2 is the sum of two prime numbers (a prime number being a number, like 3 or 111, which is divisible only by itself and 1). Though the Conjecture was stated by Goldbach in a letter to the great Swiss mathematician Leonhard Euler in 1742, it has defied all attempts by mathematicians to prove it right or wrong.

The question of whether Goldbach's Conjecture is correct can be couched as a halting problem. So too can the matter of whether or not Fermat's Last Theorem is right. 'To solve these important problems – and many others – it is therefore necessary only to know the bits of Omega!' says Calude.

Omega, it appears, is so much more than a maximally unknowable, maximally uncomputable, maximally random number. It is so much more than the distillation of all conceivable halting problems. As Chaitin puts it: 'Omega is also the diamond-hard distilled and crystallised essence of mathematical truth.'

Bennett is even more lyrical. 'Throughout history mystics and philosophers have sought a compact key to universal wisdom, a finite formula or text which, when known and understood, would provide the answer to every question,' he says. 'The Bible, the Koran, the I Ching and the medieval Jewish Cabala exemplify this hope. Omega is in many senses a cabalistic number. It can be known of, but not known through human reason. To know it in detail, one would have to accept its uncomputable digit sequence on faith, like words of a sacred text.'

There you have it. Omega is a 'compact key to universal wisdom'. It provides the answer to every question – at least, every mathematical question.

John Casti of the Technical University of Vienna goes one step farther. 'Omega's digits encode "the secret of the universe",' he says. 'Almost every unsolved problem in mathematics and many in physics and elsewhere could be settled by knowing enough digits of Omega.'*

Of course, Omega may contain the 'secret of the universe' but it is unknowable. In fact, it is worse than this. 'Even if, by some kind of miracle, we get the first 10,000 digits of Omega, the task of finding the problems whose answers are embodied in these bits is computable but unrealistically difficult,' says Calude. 'Technically, the time it takes to find all halting programs of length less than n grows faster than any computable function of n.'

In other words, we will be in the position of the characters in Adams's *The Hitchhiker's Guide to the Galaxy*. They knew the answer to Life, the Universe and Everything is forty-two. Unfortunately, the hard part was knowing the question.

* Actually, the early twentieth-century French mathematician Émile Borel was the first to show how a number could encapsulate the answers to all conceivable questions. Take the French alphabet, he said, including blanks, digits, punctuation marks, upper case, lower case, letters with accents, and everything. Then start making a list. Start off with all possible 1-character sequences, then all possible 2-character sequences, and so on. Eventually, you will get all possible successions of characters of any length, in alphabetical order. Most will of course be nonsense. Nevertheless, in the list, you will find every conceivable question in French – in fact, everything you can write in French.

Next, said Borel, number the sequences you have created. Then imagine a

Determining all the digits of Omega is clearly an impossibility for lowly human beings. The number in its entirety is really knowable only by God. Incredibly, however, Calude has managed to calculate the first 64 digits of Omega – or at least *an* Omega. Those digits are the ones shown at the beginning of this chapter.

Calude was able to calculate 64 bits of a nominally uncomputable number because, contrary to everything that has been said up to now, the computing barrier discovered by Turing can actually be broken. This is because Turing defined the halting problem for a classical Turing machine – a familiar general-purpose computer. However, nature permits types of machines that Turing did not anticipate such as 'quantum computers'. These are 'accelerated Turing machines'. It may be possible to use them to solve the halting problem and compute other apparently uncomputable things. Calude himself used ingenious mathematical tricks to partially circumvent the Turing barrier and compute 64 bits of Omega – a feat even Chaitin, the inventor of Omega, had believed impossible.

Calude's demonstration that it is possible to know at least some of the digits of Omega has left a strong impression on Chaitin. He has gone so far as to suggest that knowledge of Omega could be used to characterise the level of development of human civilisation. Chaitin points out that, in the seventeenth century, the mathematician Gottfried Leibniz

---

number $0.d_1d_2d_3 \ldots$ whose nth digit, d_n is a 1 if the nth element of the list is a valid yes/no question in French whose answer is, yes, and whose nth digit is 2 if the nth element is a valid yes/no question whose answer is, no. If the nth element of the list is garbage, not valid French, or valid French but not a yes/no question, then the nth digit is 0.

So Borel had a number that gives the answer to every yes/no question you can ask in French – about religion, about maths, about physics – and it is all in one number! Of course, such a number would contain an infinite amount of information, which would make actually ever knowing it a bit unrealistic. It would be just like Omega. In fact, Borel's number is actually related to Omega.

observed that, at any particular time, we may know all the interesting mathematical theorems with proofs of up to any given size, and that this knowledge could be used to measure human progress. 'Instead, I would propose that human progress – purely intellectual not moral – be measured by the number of bits of Omega that we have been able to determine,' says Chaitin.

'What you don't know is also a kind of knowledge,' said Jostein Gaarder in *Sophie's World*.

Calude has even suggested that, if we wish to signal our existence to the stars, the way to impress an extraterrestrial civilisation and show ourselves worthy of contact may be to broadcast as many digits as we can of Omega.

But Omega, in addition to being related to what computers can and cannot do, and distilling the answers to all mathematical questions, has yet more miraculous properties (there appear to be no end to them!). To understand what those properties are, it is necessary to appreciate a major crisis which occurred in mathematics in the late nineteenth century. The crisis was triggered by a field of mathematics called 'set theory'.

Undecidability

Set theory is concerned with groups of objects known – not surprisingly – as 'sets'. Examples include the set of all countries beginning with the letter 'A'; the set of all odd numbers; and the set of all mammals. Sets can be related to each other. For instance, one set can be contained within another set. The set of all marsupials, for instance, is a 'subset' of the set of all mammals which, in turn, is a subset of the set of all animals.

Although the basic idea of a set is straightforward, it turns out that set theory permits the existence of a particularly catastrophic set – the set of all sets that are not members of themselves. Why is this catastrophic? Well, try asking whether this set is a member of itself? Immediately, it is apparent that the set is a member of itself only if it is not a member of itself! 'It's like the paradox of the village barber who shaves every man in the village who doesn't shave himself,' says Chaitin. 'Who shaves the barber? He shaves himself if and only if he doesn't shave himself. It is a contradiction.'

This set paradox is closely related to an even older paradox: the declaration (in the sixth century BC) by the Greek philosopher Epimenides: 'This statement is false!' Since the statement is true only if it is false, it is neither true nor false. And Epimenides' paradox is in turn just another form of the 'liar's paradox': the assertion by someone: 'I am lying.'

For mathematicians of the late nineteenth century the contradiction in set theory was the stuff of nightmares. The very foundation of mathematics was logical reasoning. Yet here was a case where logical reasoning led to an absurdity. Mathematics was widely regarded as a realm of pure, clear-cut truths, a lofty ethereal domain far removed from the messiness of the everyday world. After all, it dealt with things which were true, demonstrably and beyond any possible doubt, not simply at this moment but throughout all eternity. Yet now the mathematicians' sanctuary of beauty and perfection was under mortal threat. At all costs the contradiction in set theory must be eradicated. And the man who took on the task of eradicating it was the greatest mathematician of his day, the German David Hilbert.

There are thousands of fields of mathematics, many of which are interconnected. However, each has the same basic structure. On top of a bedrock of 'axioms', mathematicians erect a scaffolding of 'theorems'. The axioms are self-evident truths, simple assertions on which all mathematicians can agree (no progress in mathematics is possible without assuming some things). The theorems are the logical consequences of the axioms. For instance, Euclidian, or flat-paper, geometry consists of a handful of axioms about straight lines and the angles between them – for example, 'parallel lines never meet' – plus the theorems that can be deduced from such axioms – for example, 'the internal angles of a triangle always add up to 180 degrees'.

Hilbert's big idea was to first identify a small group of axioms as the bedrock of all of mathematics. Once this was done, the next step was to spell out in painstaking detail the logical rules for getting from the axioms to theorems (or vice versa). This would make it possible to 'prove' any mathematical statement – that is, show that it can be obtained by logical steps from the bedrock axioms and so is a bona-fide theorem.

In short, what Hilbert had in mind was finding a proof-checking 'algorithm' – a procedure for checking that each step in a given proof is logically watertight. If mathematicians possessed such a procedure they would in theory be able to run through all possible proofs, starting with the simple ones and progressing to more complex ones; check whether they are correct; and see what theorems they led to. In this way they would be able to generate an infinite list of all provable mathematical statements – that is, all theorems.

If a mathematical statement is true, Hilbert's mindless approach would therefore eventually find the proof. If a statement cannot be proved, Hilbert's mindless method would go on for ever, unless a proof that the statement is false was found.

The mechanical nature of Hilbert's proof-checking procedure was crucially important. After all, if it could be applied mindlessly, without any need to know how mathematics worked, then it would be something absolutely everyone could agree on. Hilbert would have taken the process of doing mathematics and set it in stone. He would have removed from the subject all the ambiguities of everyday language and reasoning. There would be no room left for contradictions such as the one that appeared to have cropped up in set theory.

Hilbert did not know it – could not have known it – but the mechanical proof-checking procedure he envisaged was nothing less than a computer program running on a computer! 'How many people realise that the computer was actually conceived and born in the abstract field of pure mathematics?' asked Chaitin.

Hilbert's programme to weed out the paradoxes from mathematics was hugely ambitious. He fully expected it to take decades to carry out. But what he did not realise – and nor did anyone else – was that the programme was impossible!

In 1931, an obscure Austrian mathematician called Kurt Gödel showed that, no matter what set of axioms you select as the ultimate bedrock of all mathematics, there will always be theorems – perfectly legitimate theorems – that you can never deduce from the axioms. Contrary to all expectations, the perfect world of mathematics is plagued by 'undecidable' theorems – things which are true but which can never be proved to be true by logical, rational reasoning.

Gödel proved his result in the most ingenious way. He managed to embed in arithmetic – one of the most basic fields of mathematics – the self-referential declaration 'this statement is unprovable'. Since this required him to make a piece of arithmetic actually refer to itself, it was an immensely difficult task. However, by embedding the troublesome statement in arithmetic, Gödel had buried an atomic bomb in the very fabric of mathematics. 'This statement is unprovable' is, after all, the 'liar's paradox' in another guise. If it is true, mathematics admits false statements that cannot be proved – it is inconsistent. If it is false, it admits undecidable statements that can never be settled – it is incomplete.

Incompleteness is very bad for mathematics but inconsistency is truly terrible. False statements would be like a plague of moths gnawing at its very fabric. There was no choice for mathematicians but to accept the lesser of Gödel's two evils. Mathematics must be terminally incomplete. To everyone's profound shock, it contained theorems that could never be proved to be true.

'All theorems rest on premises,' declared Aristotle. Gödel's 'incompleteness theorem' shows that the great man was sorely mistaken. High above the mathematical bedrock there are pieces of mathematical scaffolding floating impossibly in mid-air.

The obvious way to reach these free-floating theorems is by building up the bedrock – that is, adding more axioms. However, this will not help. According to Gödel's incompleteness theorem, no matter how many axioms are added, there will always be theorems floating in the sky, perpetually out of reach. There will always be theorems that are true but that can never be proved to be true, at least by logical, rational reasoning.

To say that Gödel's discovery was deeply distressing to mathematicians is a bit of an understatement. As pointed out, mathematicians had believed mathematics was a realm of certain truths, far from the messy uncertainty of everyday life. This is precisely what had attracted many of them to the field in the first place. But, contrary to expectations, mathematics turned out to be a realm where many things are up in the air, many things are messy, many things are uncertain.*

Some mathematicians could not hide their despair at this unhappy revelation: 'Gödel's result has been a constant drain on my enthusiasm and determination,' wrote Hermann Weyl.

No matter how unpalatable it might be, however, Gödel's result was incontrovertible. Mathematicians had no choice but get used to it – even to

---

* By an odd coincidence, physicists were having a similarly shocking experience at about the same time. The microscopic realm of atoms, they discovered, was a place of unpredictability and uncertainty. Not only do things such as the disintegration of an atom occur for no reason at all, it is not even possible to be 100 per cent certain of basic matters like the position and speed of atoms. This uncertainty was quantified in Heisenberg's uncertainty principle. If Gödel dropped a bombshell in mathematics, Heisenberg can be said to have dropped one in physics.

revere it. Many now consider the publication of Gödel's incompleteness theorem to be the most significant event in the whole of twentieth-century mathematics. 'Gödel's incompleteness theorem has the same scientific status as Einstein's principle of relativity, Heisenberg's uncertainty principle and Watson and Crick's double helix model of DNA,' says Calude.

But, if things were bad in the world of mathematics after Gödel discovered incompleteness, they got a whole lot worse five years later. That was when Turing discovered uncomputability.[*] Not only did mathematics contain things that were undecidable, it also contained things such as the halting problem that were uncomputable.

Undecidability is in fact deeply connected to uncomputability.[†] Not only that but both undecidability and uncomputability are also deeply connected to Chaitin's idea that the complexity of a number is synonymous with the shortest program that can generate the number.

This is not obvious at all. However, recall that Omega is the ultimate in irreducible information. This means it cannot be generated by a program shorter than itself, which is the same as saying it cannot be compressed into a shorter string of bits than itself. Now think of one of those free-

---

[*] The 29 March 1999 issue of *Time* magazine included both Gödel and Turing among the twenty greatest scientists and thinkers of the twentieth century.

[†] In fact, undecidability is a consequence of uncomputability. If it were always possible to start with some axioms and 'prove' that a given program halts or that it never does, that will give you a way to 'compute' in advance whether a program halts or not. How? You simply run through all possible proofs, starting with the simplest ones, checking which ones are correct, until either you find a proof the program will halt eventually or you find a proof that it is never going to halt. Since Turing showed that computing in advance whether or not a program will halt is impossible, it follows that this procedure too is impossible. It follows that there must be proofs – such as the proof that a given program will halt – that cannot be found by this logical, step-by-step process. In other words, there are proofs that cannot be deduced from any conceivable axioms, and mathematics is incomplete.

floating theorems that Gödel discovered are an inevitable feature of mathematics. It cannot be reached by logical deduction from any axioms, which is the same as saying it cannot be deduced from any principles simpler than itself, compressed into any set of axioms. See the parallel?

Compressibility and What Scientists Do

Here is another interesting thing. Reducing theorems to a small number of axioms turns out to be deeply reminiscent of what scientists do. The mark of a good scientific theory, after all, is that it describes a large number of observations of the world while making only a small number of assumptions. In the words of the Nobel Prize-winning American physicist Richard Feynman: 'When you get it right, it is obvious it is right – at least if you have any experience – because usually what happens is that more comes out than goes in.'

Chaitin, as the inventor of AIT, has a unique take on this idea. 'A scientific theory is really just a computer program that calculates observations,' he says. 'The smaller and more concise the computer program the better the theory.'

Scientists have long known that, if there are two competing theories both of which explain the same phenomenon, the one that makes the least assumptions is invariably the true one. This rule of thumb, known as 'Ockham's razor', was first noted by William of Ockham, a Franciscan friar living in the fourteenth century.

By the criterion of Ockham's razor, for instance, Creationism is inferior to the scientific view of the origin of the Universe because it requires many more assumptions. What you get out is not much better than what you put in. As Leibniz observed more than three centuries ago, a theory is convincing only to the extent that it is substantially simpler than the facts it attempts to explain.

According to Chaitin, AIT puts Ockham's razor on a precise footing for the first time. 'Understanding is compression,' says Chaitin. 'Ockham's razor is simply saying that the best scientific theory is the most compressible.' He amplifies this. 'A concise computer program for calculating something is like an elegant theory that explains some phenomenon,' says Chaitin. 'And if no concise theory exists, the phenomenon has no explanation, no pattern, there is nothing interesting about it – it is just what it is, that's all.'

In physics, the Holy Grail is a 'Theory of Everything', which distils all the fundamental features of reality into one simple set of equations that could be written on the back of a postage stamp. 'From the point of view of AIT, the search for the Theory of Everything is the quest for an ultimate compression of the world,' says Chaitin.

'The most incomprehensible thing about the universe,' Einstein famously said, 'is that it is comprehensible.' Chaitin, who equates comprehension with compression, would rephrase this. The most incomprehensible thing about the Universe is that it is compressible. This feature of the world is the reason we have been able to divine universal laws of nature, which apply in all places and at all times – laws which have enabled us to build computers and nuclear reactors and gain some degree of mastery over nature.

To Chaitin the compressibility of the Universe is a wonder. 'For some reason, God employed the least amount of stuff to build the world,' he says. 'For some reason, the laws of physics are as simple and beautiful as they can be and allow us, intelligent beings, to evolve.' This is a modern version of something noted by Leibniz: 'God has chosen the most perfect world,' he wrote. 'The one which is the most simple in hypotheses and the most rich in phenomena.'

Though we do not know why the laws underpinning the Universe are simple, the faith that they are is a powerful driving force of science. According to Feynman: 'It is possible to know when you are right way ahead of checking all the consequences. Truth is recognisable by its beauty and simplicity.'

Randomness

Back to Gödel. Although he had shocked and depressed mathematicians by showing that mathematics contains theorems which are undecidable, surprisingly his result did not make any difference to the day-to-day doing of mathematics. Weyl's pessimism was misplaced. 'Mathematicians, in their everyday work, simply do not come across results that state that they themselves are unprovable,' says Chaitin. 'Consequently, the places in mathematics where you get into trouble seem too remote, too strange, too atypical to matter.'

A more serious worry was Turing's demonstration that there are things in the world which were completely uncomputable. This is a very concrete result. It refers, after all, to computer programs, which actually calculate things. On the other hand, the program Turing considered merely tried to figure out whether another program halts or not. It is hardly typical of today's computer programs, which carry out word processing or surf the Internet. Not surprisingly, therefore, none of these programs turns out to be undermined in any discernible way by the uncomputability of the halting problem.

It would seem that uncomputability and undecidability are too esoteric to bother about, that they can be swept under the carpet and safely forgotten about. This is indeed how it appeared for a long while.

All was tranquil and quiet in the garden of pure mathematics. But then the gate squeaked on its rusty hinges and in walked Chaitin.

From the time he had been a teenager, Chaitin had been convinced that Gödel and Turing's results had far more serious implications for mathematics than anyone guessed. And he had resolved to find out what those implications were. It was this quest that had led him to invent AIT.

AIT is of course founded on the idea that the complexity, or information content, of a number is synonymous with the shortest computer program that can generate the number. However, at the core of AIT – just like at the core of Turing and Gödel's work – there is a paradox. It is actually impossible to ever be sure you have found the shortest possible program!*

A shortest program, of course, does exist. But this is not the point. The point is that, although it exists, you can never be sure you have found it. Determining whether you have turns out to be an uncomputable problem.

AIT is founded on uncomputability. The whole field is as riddled with holes as a Swiss cheese. Uncomputability in fact follows from AIT.†
And so does Gödel's incompleteness theorem. This turns out to be equivalent to the fact that it is impossible to prove that a sequence of

---

* Say there is a program that can decide whether a given program, p, is the shortest possible program capable of producing a given output. Now consider a program, P, whose output is the output of the smallest program, p, bigger than P that is capable of producing the given output. But P is too small a program to produce the same output as p. There is a contradiction! Therefore, an algorithm for deciding if a program p is as small as possible cannot exist.

† If you could always decide in advance whether a program halts or not, you could systematically check whether each small program halts or not, and if it does halt, run it and see what it computes, until you find the smallest program that computes a given number. But this would contradict Chaitin's result that you cannot ever be sure you have the smallest program for generating a given number. Consequently, there can be no general solution to the halting problem. It is uncomputable.

digits is incompressible – that is, the shortest program has been found. 'Everywhere you turn in my theory you find incompleteness,' says Chaitin. 'Why? Because the very first question you ask in my theory gets you into trouble. You measure the complexity of something by the size of the smallest computer program for calculating it. But how can you be sure that what you have is the smallest computer program possible? The answer is that you can't!'

In Chaitin's AIT, undecidability and uncomputability take centre stage. Most mathematical problems turn out to be uncomputable. Most mathematical questions are not, even in principle, decidable. 'Incompleteness doesn't just happen in very unusual, pathological circumstances, as many people believed,' says Chaitin. 'My discovery is that its tendrils are everywhere.'

In mathematics, the usual assumption is that, if something is true, it is true for a reason. The reason something is true is called a proof, and the object of mathematics is to find proofs, to find the reason things are true. But the bits of Omega – AIT's crowning jewel – are random. Omega cannot be reduced to anything smaller than itself. Its 0s and 1s are like mathematical theorems that cannot be reduced or compressed down to simpler axioms. They are like bits of scaffolding floating in mid-air high above the axiomatic bedrock. They are like theorems which are true for no reason, true entirely by accident. They are random truths. 'I have shown that God not only plays dice in physics but even in pure mathematics!' says Chaitin.[*]

Chaitin has shown that Gödel's and Turing's results were just the tip of

---

[*] This is a reference to Einstein. Appalled by quantum theory, which maintained that the world of atoms was ruled by random chance, he said: 'God does not play dice with the universe.' Unfortunately, he was wrong. As Stephen Hawking has wryly pointed out: 'Not only does God play dice, he throws them where we cannot see them.'

the iceberg. Most of mathematics is composed of random truths. 'In a nutshell, Gödel discovered incompleteness, Turing discovered uncomputability, and I discovered randomness – that's the amazing fact that some mathematical statements are true for no reason, they're true by accident,' says Chaitin.

Randomness is the key new idea. 'Randomness is where reason stops, it's a statement that things are accidental, meaningless, unpredictable and happen for no reason,' says Chaitin.

Chaitin has even found places where randomness crops up in the very foundation of pure mathematics – 'number theory'. 'If randomness is even in something as basic as number theory, where else is it?' says Chaitin. 'My hunch is it's everywhere.'

Chaitin sees the mathematics which mathematicians have discovered so far as confined to a chain of small islands. On each of the islands are provable truths, the things which are true for a reason. For instance, on one island there are algebraic truths and arithmetic truths and calculus. And everything on each island is connected to everything else by threads of logic so it is possible to get from one thing to another simply by applying reason. However, the island chain is lost in an unimaginably vast ocean. The ocean is filled with random truths, theorems disconnected for ever from everything else, tiny 'atoms' of mathematical truth.

Chaitin thinks that Goldbach's Conjecture, which has stubbornly defied all attempts to prove it true or false, may be just such a random truth. We just happened to have stumbled on it by accident. If he is right, it will never be proved right or wrong. There will be no way to deduce it from any conceivable set of axioms. Sooner or later, in fact, the Goldbach conjecture will have to be accepted as a shiny new axiom in its own right, a tiny atom plucked from the vast ocean of random truths.

In this context, Calude asks an intriguing question: 'Is the existence of God an axiom or a theorem?'

Chaitin is saying that the mathematical universe has infinite complexity and is therefore not fully comprehensible to human beings. 'There's this vast world of mathematical truth out there – an infinite amount of information – but any given set of axioms only captures a tiny, finite amount of this information,' says Chaitin. 'This, in a nutshell, is why Gödel's incompleteness is natural and inevitable rather than mysterious and complicated.'

Not surprisingly, the idea that, in some areas of mathematics, mathematical truth is completely random, unstructured, patternless and incomprehensible, is deeply upsetting to mathematicians. Some might close their eyes, view randomness as a cancer eating away at mathematics which they would rather not look at, but Chaitin thinks that it is about time people opened their eyes. And rather than seeing it as bad, he sees it as good. 'Randomness is the real foundation of mathematics,' he says. 'It is a revolutionary change in our worldview.'

This is explosive stuff. But Chaitin is able to risk the ire of the mathematical community because he is an outsider. He works for IBM at its Thomas J. Watson Research Center in Yorktown Heights, New York. In fact, he helped to develop the company's influential 'Unix' work station, the IBM RS/6000. Chaitin does not believe it is possible to break the mould from within mathematics. 'To be a revolutionary it may be necessary to be on the outside,' he says.

Crucially, the language of physics turns out to be mathematics. The equations which describe things such as the motion of baseballs through the air and planets orbiting the Sun are mathematical equations. Many have remarked on the unreasonable effectiveness of mathematics in physics. But, if most truths in mathematics are random truths, things which are true for no reason at all, what does this say about truths in physics? 'Ultimately, can the universe be comprehended – the physical universe as well as the universe of mathematical experience?' asks Chaitin.

It all depends on whether the physical universe, like the mathematical universe, is infinitely complex. If, as most physicists believe, the world of atoms is ruled by naked chance, the Universe does indeed contain randomness. Consequently, it is infinitely complex and unknowable in its entirety by human beings.*

What this means is that physicists like Stephen Hawking, who fully expect the discovery of a Theory of Everything in the next decade or so, are destined to be disappointed. Though we may acquire such a theory, we will never know for sure whether we have the ultimate Theory of Everything. We will never be able to prove the compression to be the ultimate one. There will always be the possibility that there might be a yet deeper and simpler theory, with the same explanatory power, out there waiting to be found. As the American physicist John Wheeler, famous for coining the term 'black hole', has pointed out: 'Even if physicists one day get their hands on a Theory of Everything, they will still face the unanswerable question: why does nature obey this set of equations and not another?'

The Anglo-American physicist Freeman Dyson sees the impossibility of finding a Theory of Everything as a good thing. Unlike the pessimistic Weyl, who described incompleteness as a constant drain on his enthusiasm, Dyson sees it as an insurance policy that science will go on for ever. Though a Theory of Everything may be elusive, the mundane, day-to-day practice of doing physics will go on. The discoveries of Gödel and Turing do not appear to limit them in any way.

---

* It is possible, however, that the randomness in the Universe is only pseudo-randomness. This is the controversial view of Stephen Wolfram (*see* Chapter 2, 'Cosmic Computer'). He believes that the consequences of the simple laws of physics are extremely complicated and so the Universe is only simulating randomness. If he is right, then the Universe is ultimately comprehensible. It would not follow that, because the laws of physics are mathematical and mathematics is incomplete, that physics is incomplete too.

As Chaitin points out, we have no trouble building and operating computers, far and away the most complicated machines ever constructed. 'Here we have a case where the physical behaviour of a big chunk of the universe is very understandable and very predictable and follows definite laws,' says Chaitin.

Chaitin's insights raise a fascinating and intriguing question about how exactly mathematicians actually do mathematics, how they find new theorems. While step-by-step reasoning and logic enables them to move from one idea to another idea within an island in the great ocean of mathematical theorems, it does not allow them to get from island to island. But this is something they emphatically *do* do.

Reason and logic are insufficient. Chaitin therefore thinks that mathematicians discover mathematics using insights which go beyond reason and logic. He thinks they use the kind of flashes of inspiration and intuition artists talk about. 'Mathematics isn't about the consequences of rules, it's about creativity and imagination,' says Chaitin. 'For this reason it is possible to argue that the incompleteness theorems do not limit what mathematicians do.'

'Mathematical proof is an essentially creative activity,' said Emil Post, who came up with the idea of a Turing machine independently of Turing. The brain is doing something more than any mere computer. Many people have suggested this before. 'You can know much more than you can ever prove,' said Feynman. But who would have thought the idea would receive support from a field as abstract and esoteric as pure mathematics?

7
Patterns in the Void

Where do the laws of physics come from?
They are the laws of nothing!

What immortal hand or eye,
Could frame thy fearful symmetry?
William Blake, 'The Tiger'

Nothing will come of nothing
William Shakespeare, *King Lear*

Where do the laws that govern our Universe come from? It is one of the most basic questions that can be asked about the Universe. The answer, however, may be ridiculously trivial, according to an American physicist. 'The laws of physics are simply the laws of nothing,' says Victor Stenger of the University of Hawaii. 'Put another way, they are precisely the laws that would reign in a featureless void, utterly empty of matter, energy or anything else.'

If Stenger, a retired experimental particle physicist, is right, it may shed light on arguably the ultimate cosmic question: How did something come from nothing? 'If the laws of physics are the same laws as the laws of an empty void, the transition from nothing to something may not have been as difficult as people have assumed,' says Stenger. 'Our

Universe may be no more than re-arranged, re-structured nothingness.' This is highly provocative. To understand how Stenger has arrived at such a view, it is first necessary to understand an idea which has proved a powerful guiding light to physicists groping for understanding of the Universe. That idea is 'symmetry'.

Symmetry is about the properties of an object that do not change if something is done to it. Take, for instance, a starfish. If it is rotated a fifth of a complete turn, it looks the same. Mathematicians say that a starfish is symmetric with respect to rotations about its centre – or, more precisely, with respect to rotations of a fifth of a complete turn about its centre. 'On the face of it, symmetry appears to have nothing whatsoever to do with physics,' says Stenger. 'However, it turns out to be crucial.'

The surprising connection between physics and symmetry was discovered by a female German mathematician in 1918. In a remarkable paper, which Einstein among others greatly admired, Emmy Noether showed that some of the fundamental laws of physics are nothing more than the consequences of simple symmetries of space and time.

Take, for instance, the fact that the laws of physics are the same today as they were yesterday. If this were not the case and, say, the law of gravity were different yesterday, you might find that your weight has overnight gone from 70 kilograms to 150 kilograms. Because the laws of physics do not change with time, you can be sure that if you do a particular experiment on a Tuesday or on Wednesday, or any other day of the week, you will get the same result. Mathematicians say that the laws of physics are symmetric with respect to 'translations in time'.

It is not obvious at all that a law of physics follows from this. However, Noether's surprising discovery was that it does. Furthermore, it is one of the cornerstones of physics – the law of conservation of energy, which states that energy can never be created or destroyed, merely changed from one form into another.

And the law of conservation of energy turns out to be not the only fundamental law of physics intimately connected to symmetry. Take the fact that the laws of physics are the same in one location as in another – for instance, in New York and in London. Mathematicians say that the laws of physics are symmetric with respect to 'translations in space'. Noether discovered that this symmetry leads to another great cornerstone of physics – the law of conservation of momentum. This states that a quantity called 'momentum' can never be created or destroyed. The momentum of a body – the product of its mass and its velocity – is in fact a measure of how difficult it is to stop the body moving. A slow-moving elephant has a lot of momentum and a fast-moving pigeon not nearly so much.

Another connection between the laws of physics and symmetry stems from the fact that the laws of physics are the same in any orientation. In other words, it does not matter, when you carry out an experiment, whether you align your apparatus north–south, east–west or along any other direction. Mathematicians say the laws of physics are symmetric with respect to 'rotations in space'. Noether discovered that this symmetry leads to the law of conservation of angular momentum. This states that a quantity known as 'angular momentum' – a measure of how difficult it is to stop a rotating body – cannot be created or destroyed.

The $64,000 question is, of course, why should there be any connection between symmetry and the laws of physics? Why does the fact that the laws of physics are the same yesterday and today lead to the law of conservation of energy? Why does the fact that the laws of physics are the same in London and in New York lead to the conservation of momentum? And why does the fact that they are the same in any orientation lead to the conservation of angular momentum?

Non-mathematically, the connection between physics and symmetry is not easy to see – which is of course why nobody noticed it before

Noether in 1918. 'Nevertheless, it is possible to get a hint of why there is a connection,' says Stenger. 'Imagine a car travelling along a road, which is featureless apart from telegraph poles spaced at regular intervals. Imagine also that the car's speed is constant and a movie is taken of the car.'

According to Stenger, it will be impossible to determine simply from watching the movie where exactly the car is along the road. After all, every telegraph pole will look exactly like every other telegraph pole. This shows that the inability to distinguish one location of the car from another – translational symmetry – is in some way connected with the fact that the speed, or momentum, of the car is not changing. If, however, the car's momentum does change – if it brakes to a sudden halt – the telegraph poles are no longer indistinguishable. One is singled out as special – the one nearest to where the car stops – breaking the translational symmetry. 'The example of the car hopefully hints that there is a deep connection between symmetry and the laws of physics,' says Stenger.

The laws that Noether was able to show were mere consequences of symmetries are all symmetries of either space on its own or of time on its own. But is there a symmetry of space and time together? The answer is an emphatic *yes*. The symmetry leads to one of the greatest ideas in the history of science – Einstein's 'special' theory of relativity.

Special Relativity

The laws of physics turn out not only to be the same in different orientations of space but also in different orientations of space and time together. This takes a bit of explaining. After all, how is it possible to rotate something in time? Well, it is only possible to do so if time is a dimension much like a dimension of space. This, it turns out, is exactly what Einstein discovered in 1905.

According to Einstein, the space and time dimensions are so similar that intervals of time can in fact change into intervals of space and vice versa. The reason is that we have been hoodwinked into believing that space and time are actually fundamental things. In fact, they are merely artefacts of our particular viewpoint. One person's interval of space is not the same as another person's interval of space. And the same goes for time.

Imagine a stick suspended from the ceiling of a square shed so that it can swing about horizontally just like a giant compass needle. And say we can view the stick only through a window in one wall of the shed and a window in the adjacent wall. Furthermore, it is gloomy in the room so we cannot actually tell that it is a suspended stick, just an 'object' with a certain extent when we view it through one window and a certain extent when we view it through the other window. We might decide to call these its 'length' and 'width'.

Now, say the shed is on a turntable and is rotating about the object (this is a very odd shed!) – as we look through the windows, we will see the width gradually turn into the length and vice versa. Gradually, the truth will dawn on us. We were stupid to describe the object in terms of its 'width' and 'length'. These are not fundamental things at all but merely artefacts of our viewpoint.

So it is with space and time. In reality, there is a thing called space-time – a seamless blending together of space and time. Just as the way we divide the extent of the stick into width and length depends on our viewpoint, the manner in which we split space–time into space and time depends on our viewpoint. The only difference is that, in the case of space–time, what determines our particular view of it is not anything as crude as a window we are looking through but how fast we are moving.

In fact, as pointed out earlier, you have to be travelling close to the speed of light to appreciably change your viewpoint of space and time. The speed of light, however, is roughly a million times faster than a passenger jet. Consequently, we never see space turn into time and time into space. We have been hoodwinked into believing space and time are fundamental because we live our lives in the cosmic slow lane. That is why it took the genius of Einstein to see the truth. Space and time are but projections of the seamless entity of space-time. They are like shadows on a wall. Space-time is the fundamental thing.

In practice, intervals of space are measured with a ruler and intervals of time with a clock. Rulers and clocks therefore look different depending on your viewpoint – how fast you are moving past them. More precisely, as pointed out before, moving rulers shrink and moving clocks slow down. It makes not the slightest difference whether you think it is the clocks and rulers that are moving or you. As Einstein realised, the only motion that is in any way meaningful is 'relative' motion.

The precise way space 'contracts' and time 'dilates' for a moving observer is given by a mathematical formula called the 'Lorentz transformation'. Einstein postulated that all observers moving at uniform speed have equally valid viewpoints. Consequently, the laws of physics must be the same when space and time are subject to a Lorentz transformation. In the jargon, they are symmetric under a Lorentz transformation. Here we come back to Noether. 'The existence of a symmetry, as Noether discovered, always implies a law of physics,' says Stenger. 'In this case, the law turns out to be one not often considered to be a law – the law of the conservation of the speed of light.'

This says that the speed of light appears exactly the same for everyone in uniform motion – that is, travelling at constant speed – no matter what their speed. In fact, the Lorentz transformation is simply a recipe which tells everyone what must happen to their intervals of space and intervals of time so that they all measure precisely the same speed for a beam of light.*

Because of the conservation of the speed of light, whether a light beam is coming towards you or going away from you, it will always appear to be travelling at the same speed. Contrast this with a motorbike that is travelling at 100 kilometres per hour on a road where you are travelling in a car also at 100 kilometres per hour. If the motorbike is travelling your way, it will appear to be standing still; if it is coming towards you, it will appear to be travelling at 200 kilometres per hour. Light, in marked contrast to the motorbike, appears to everyone, no matter what their motion, to travel at the same speed. It really is the rock on which the Universe is built, with space and time but shifting sand.

No doubt this all seems very complicated. We are talking, after all, about Einstein's special theory of relativity. However, everything Einstein discovered is connected to a very simple symmetry of space-time. Specifically, a Lorentz transformation, which slows time and shrinks space for a moving observer, turns out to be nothing more than a rotation in space-time.

Recall that time is actually a dimension much like the three dimensions of space – north–south, east–west and up–down. A total of four dimensions is impossible to imagine. However, if two space

---

* Historically, physicists came to this the other way round. Einstein postulated that the laws of motion must appear the same to all observers moving at constant speed relative to each other. This was something Galileo had noticed – a ball dropped to the deck of a moving ship fell in exactly the same way as a ball dropped on stationary dry land, so the laws of motion are the same whatever your velocity. Einstein, in addition, postulated that the laws of electricity and magnetism – electromagnetism – must also appear the same to all observers. Because it had been discovered by James Clerk Maxwell that light was actually an 'electromagnetic' wave, this meant that the speed of light must appear the same to everyone irrespective of their speed. These considerations led Einstein to the Lorentz transformation.

dimensions are conveniently ignored, space-time can be imagined as a two-dimensional map with, say, time going up the page and space going across the page.

In a real map, the location of, say, a castle, might be specified as 5 kilometres north of a certain town and 0 kilometres east. Now, say you rotate the north–south and east–west directions so that north is now off at some angle to the vertical. The location of the castle might now be 3 kilometres from the town in the new north direction and 4 kilometres in the new east direction. In the new 'coordinate system', some of the old east–west coordinate has changed into new north–south coordinate, and vice versa.

Well, in a space-time map, something similar happens. If the space and time directions are rotated, some of what was time becomes space and vice versa. And this is entirely equivalent to a Lorentz transformation. So, saying that the laws of physics – and, specifically, the speed of light – are unchanged under a Lorentz transformation is equivalent to saying the laws are the same no matter what the orientation of the directions of space and time.[*]

There is actually nothing deep or significant about this. If, rather than describing the location of a castle on a map one way, we described its location another way, it would be surprising if it changed anything essential about the castle. 'All Einstein did was extend this straightforward reasoning to space-time,' says Stenger. 'He simply said that, no matter how we view an object in space-time, its essential features should remain the same.'

---

[*] There is a slight twist here. Time and space dimensions are subtly different from each other, as we might have expected from our everyday experience – we can move in any direction in space but are more restricted in time. So, in all this talk about rotations of the space and time directions, the time dimension is actually 'imaginary', which has a very specific mathematical meaning.

In fact, this goes to the heart of what all the symmetries of space-time are about. They are about making sure that the fundamental laws of physics are the same for everyone, no matter what their viewpoint.

The fact that the laws of physics are symmetric under translations of space means that, if we did an experiment in London and referred all our measurements to New York – which in practice would mean adding about 5,000 kilometres to any measurements made in the direction joining London to New York and would not be a sensible thing to do – we would get the same result. Similarly, the fact that the laws of physics are symmetric under translations in time means that, if we did an experiment and referred all our time measurements to the time the dinosaurs vanished 65 million years ago, it would not affect the outcome. Also, the fact that the laws of physics are symmetric under rotations in space-time means that if we made all of our measurements using a coordinate system rotated in space-time – entirely equivalent to a coordinate system travelling through space at constant velocity – that would make no difference either.

General Relativity

There are clearly situations, however, when the laws of physics do depend on your viewpoint. Say you are in a plane which is changing its speed, or 'accelerating'. Drop a ball in the aisle and it will not fall vertically to the ground as it does in a plane flying at constant speed. The greater the acceleration, as indicated by how hard you are pushed back in your seat, the farther from the vertical the ball will fall. Obviously, the laws of physics do not appear the same from the viewpoint of two people undergoing different accelerations.

Nevertheless, the desire to write down the laws of physics in a way that does not depend on viewpoint is a powerful one. How, after all, could physicists hold up their heads if their 'universal' laws were different in different situations – for instance, for people undergoing different accelerations?

For a decade after his discovery of the special theory of relativity, Einstein struggled to find a way to write down the laws of motion in a way that is the same for everyone, no matter what their acceleration. In 1915, at the height of the First World War, he succeeded. He could do it on one condition. The condition was that a force with a very particular behaviour be added to region of space which is undergoing acceleration. The force is gravity!

Gravity, according to Einstein, is therefore not a real force at all. It is a force we invent merely so that the laws of physics still make sense to people accelerating at different rates. In fact, gravity is like other 'fictitious' forces such as 'centrifugal force'. If you are in car rounding a sharp bend, you will find yourself flung outwards. So that the laws of physics – for instance, the laws that describe how a ball dropped in the car falls to the floor – still make sense, it is necessary to invent an outward force called 'centrifugal force'. No such force exists, however, as is obvious to anyone observing you from outside the car. They can see that you are not flung outwards towards the car door. Rather, it is the car door that comes towards you because the car is following a curved path and you are continuing to move in a straight line under your 'inertia'.

The Earth orbiting the Sun is like you in the car going round the bend. It is not being constrained to circle the Sun by the 'force' of gravity. It is simply flying through space in a 'straight' line because of its inertia. Of course, it does not look like the Earth is moving in a straight line. But that is just because we are using the wrong definition of a straight line!

It was Einstein's genius to realise that space-time is actually warped by the presence of a mass like the Sun. As mentioned before, we cannot see this warpage because, as lowly three-dimensional creatures, we cannot directly experience something four-dimensional. But, just as a cannon ball on a trampoline creates a valley in the fabric of the trampoline, the Sun creates a valley in the fabric of space-time around it. And, on a warped surface, a straight line – that is, the shortest distance between two points – is a curve. This is why the Earth, which is doing nothing more than travelling under its own inertia along the shortest path possible through warped space-time, is circling the Sun.

The theory in which Einstein was able to write down the laws of physics in a so-called covariant form that was the same for everyone is called the general theory of relativity. The 'general' denotes that it applies to people no matter what their state of motion, not just to 'special' people who happen to be in uniform motion with respect to each other.

Quantum Theory

Noether made her discovery of the significance of symmetry to physics in the field of 'dynamics', which deals with the way in which bodies move through space. From symmetries of space and time separately, she was able to deduce great conservation laws such as the conservation of energy. It later became apparent to physicists that Einstein's special theory of relativity was also a consequence of a symmetry of space-time, but a symmetry of space and time combined. And symmetry proved its worth also in the general theory of relativity. To restore symmetry and make sure everyone, no matter what their motion, experienced the same laws of physics, it was necessary to invent a force – the force of gravity.

But Noether's discovery that symmetry underpins the world turns out to be an enormously fruitful idea with far-reaching implications for physics. The reason is that symmetries of space and time are not the only symmetries that spawn laws of physics. There are other, far more abstract, symmetries. 'And it is nature's most abstract symmetries that have the most profound consequences – especially in the field of quantum theory,' says Stenger.

While general relativity describes the behaviour of big things like stars and the whole Universe, quantum theory generally describes the behaviour of small things like atoms and their constituents. The theory's founding idea can be stated in a sentence: in the microscopic world of atoms and their constituents, particles behave like waves and waves like particles. The behaviour of an atom or an electron is therefore described by a wave – actually, an abstract, mathematical wave but a wave with concrete consequences nonetheless. How this quantum wave spreads through space is in turn described by the 'Schrödinger equation'.[*]

As mentioned before, the quantum wave is not in fact observable directly. The only thing with real, 'physical' significance is the square of the wave height at any point in space. This turns out to be the 'probability' of finding the particle at that point. It can be anything between 0 per cent and 100 per cent.

In practice, what all this schizophrenic wave particleness means is that a particle like an atom can do all the things that a wave can do. It is a well-known property of waves, for instance, that, if two waves can exist – say, a big wave spreading on a pond and a small ripple – then a combination of the two can also exist – a big wave with a small ripple superimposed on it. This has dramatic implications in the world of

---

[*] Physicists call the quantum wave the 'wave function'.

atoms. If there is a quantum wave with a high probability of an atom being in one place and another quantum wave with a high probability of the atom being in another place, there can – as noted earlier – be a superposition of the two waves, which corresponds to the atom being in two places at once (*see* Chapter 4, 'Keeping It Real').

And this is just one of the countless peculiarities of the quantum world.

Back to symmetry. Think of the quantum wave again. It turns out that there is an important subtlety when it comes to the wave height. It is described not by an ordinary number but by a 'complex number'. Such a number has a 'real' part – just an ordinary number – and an 'imaginary' part. It is not crucial here to know what an imaginary number is.* The key thing is simply that a complex number has two components. This means it can be visualised on a two-dimensional map.

Think of the north–south direction as the real direction and the east–west direction as the imaginary direction. Any complex number can be represented by a dot somewhere on the map. Picture a line drawn to the dot from the map's centre, or 'origin' – the spot where the north–south and east–west directions meet. It has an extent in the real direction and an extent in the imaginary direction. The line is therefore an equivalent way of representing the complex number. Think of it as an arrow pointing out from the origin.

Now, recall that at any point in space the height of the quantum wave squared is the probability of finding the particle at that particular point. Well, the square of the wave height is just the square of the length of the

---

* An imaginary number is some multiple of the square root of a negative number such as -1. Such a square root does not exist, hence the term 'imaginary'! Nevertheless, the square of an imaginary number does exist. And, because it exists, it is possible to manipulate imaginary numbers and still finish up with their squares – numbers that do exist. Physicists have found it extraordinarily fruitful to represent many quantities in nature by complex numbers – a combination of real and imaginary numbers.

arrow. So, as long as the arrow stays the same length, the probability will be unchanged. Here we are getting to the point. The length of the arrow stays the same if it is rotated about the origin, no matter what direction it ends up pointing. For technical reasons which are not worth going into, physicists call such a rotation a 'phase shift'.

What all this means is that, if the arrow representing the height of the quantum wave of a particle at each point in space is rotated by the same amount, it makes no practical difference to the particle. Physicists say the quantum wave is symmetric with respect to 'rotations in complex space' or, equivalently, symmetric to a 'global' phase change. In fact, just to confuse matters with even more jargon, they have a special phrase for this symmetry. They call it a global 'gauge' symmetry.

A symmetry, as Noether discovered, implies a law of physics. So what law does this symmetry lead to? The answer turns out to be the law of conservation of electric charge.

Electric charge comes in two types – positive or negative. Electrons, for instance, carry a negative charge and protons a positive charge. Particles with the same charge repel each other while particles with opposite charges attract each other. It has long been known that electric charge, like energy and momentum, can never be created or destroyed. But why? Noether insight provides the answer. Ultimately, it is down to an abstract symmetry – the fact that the quantum wave is symmetric with respect to global rotations in complex space.

Now you see the power of Noether's idea. Even the most ridiculously abstract symmetries – not even of anything concrete like space and time – spawn laws of physics. But the abstract symmetries of the quantum wave do not end here. There is another, very surprising, symmetry which has profound implications for physics.

Say the arrow representing the height of the quantum wave of a particle is rotated by a different amount at each point in space. The probability of finding the particle at each point in space will remain the same. Nevertheless, the laws of physics – specifically, the Schrödinger equation that governs the motion of a quantum particle – will not be the same. The reason is that the Schrödinger equation permits the quantum wave to mingle, or 'interfere', with itself. This just means that, at places where crests of one component of the quantum wave coincide with crests of another, the wave is boosted and, at places where crests coincide with troughs, it is snuffed out.

Crucially, it is the direction of the arrow at any point that tells us whether there is a peak or a trough in the quantum wave (or any possibility between). So, if the arrows are rotated by different amounts at different points in space, it will affect how the quantum wave interferes with itself. Consequently, the quantum wave will not be symmetric under 'local' rotations in complex space, or equivalently to local phase changes.

But let us not be too hasty. What will it take to restore the symmetry? Well, clearly the quantum wave will have to be rotated by a different amount at each point so that the relative directions of all the arrows are the same as they were before. This seems a very tall order. However, remarkably, there is a way to do it. Recall the way in which Einstein restored the symmetry in his general theory of relativity. He simply added a 'field' of force everywhere in space – gravity. Well, in this quantum case the symmetry can also be restored by adding a field of force. That field of force is the electromagnetic field.

The electromagnetic field is responsible for all electrical and magnetic phenomena and, in fact, glues together all normal matter, including the atoms in your body. In 1873, the Scottish physicist James Clerk Maxwell summarised all the properties of electricity and magnetism in his laws of electromagnetism. It was a tour de force of nineteenth-century physics. Yet it turns out that everything Maxwell struggled to understand for more than a decade is merely the consequence of a simple symmetry principle: the laws of motion for a quantum particle must be the same under arbitrary rotations in complex space at different points in space. According to modern-day physicists, the electromagnetic field exists for one reason, and one reason only – to restore local 'gauge' symmetry.

The remarkable fact that laws of physics respect local gauge symmetry has proved a powerful way of finding new laws of physics. A local gauge symmetry, for instance, always requires the existence of a 'field' to maintain the symmetry. An example of a gauge field is the 'strong' field of force that glues together quarks. The field is exactly what is required to restore gauge symmetry. In turn, new particles are manifestations of the fields, localised 'knots' in the fields, if you like – in the case of the strong field, 'gluons'.

Broken Symmetries

Symmetries may abound in nature. However, one thing is clear when we look around at the Universe. It does not look very symmetric. Far from it. One place in our Solar System is not the same as another place. For instance, the region of space currently occupied by Jupiter is not the same as the location of space 10 million kilometres from Jupiter along the line joining Jupiter to the Sun. Whereas the former place contains a giant ball of swirling gas 1,000 times bigger than the Earth, the latter contains only a few shreds of hydrogen gas.

This illustrates an important point. The underlying laws of physics may be symmetric. However, the detailed consequences of these laws need not be.

Think of a pencil perfectly balanced on its point. In this situation, the law of gravity is perfectly symmetric. There is a vertical force – down towards the centre of the Earth – but there is no sideways force whatsoever. The sideways force is therefore the same in all directions – in the jargon, it is symmetric with respect to orientation in space.

But, as we all know, a pencil balanced on its tip is not stable. It is constantly being buffeted by air molecules and the slightest imbalance in that buffeting is enough to cause the pencil to fall. It might fall in the direction of north or east or in any direction at all. It does not matter. The point is (no pun intended!) that one direction will turn out to be special – the direction in which the pencil falls.

Here, we see that, although the underlying law of gravity is symmetric with respect to orientation in space, the consequence of that law is not. Physicists talk of the symmetry 'spontaneously breaking'.

As it is for pencils, so it is for the Universe. Physicists have a strong belief that the fundamental laws of the Universe are symmetric but that, in creating the world, the symmetry has been spontaneously broken.

In general, symmetries spontaneously break as the temperature drops. Take water. The liquid looks the same at every point and in every direction too. It is symmetric, in other words. Cool it until it freezes, however, and the molecules line themselves up regularly in a 'crystal lattice' in which certain directions are distinguishable from all others. What is more, cracks and fissures may spontaneously appear. Things no longer look the same at every place and the same in every direction. Ice is less symmetric than water. Cooling the water caused the symmetry to spontaneously break.

The significance of this for the Universe is profound. The Universe began almost 13.7 billion years ago in a dense, hot state – the Big Bang – and has been expanding and cooling ever since. Because the Universe was hotter in the past, we would expect it to have been more symmetric. And, in fact, our experiments prove this to have been the case.

Take two of nature's four fundamental forces – the electromagnetic force and the weak force. These have a very different character. For instance, the electromagnetic force has an infinite range whereas the weak force has an extremely short range. This difference can be traced back to the 'carriers' of the two forces.

The carrier of the electromagnetic force is the photon and the carriers of the weak force are the W-, W+ and Z0 bosons. The 'range' of a force turns out to be inversely related to the mass of the force carriers. It is because the photon has no mass that the range of the electromagnetic force is infinite. And it is because the three 'vector bosons' have large masses – around about 100 times the mass of a proton – that the weak force has a very short range.

The situation with the electromagnetic and weak forces is very messy and unsymmetric. However, the belief among physicists is that, early in the Universe, when the temperature was much higher, things were very different. The vector bosons, instead of having large masses, had no mass, just like the photon. The two forces were identical. They behaved as a single 'electroweak force'. Spontaneous symmetry breaking turned a single massless force-carrier into four very different force-carrying particles. Sure enough, physicists confirmed this scenario in the early 1980s when they recreated the temperature of the early Universe at CERN, the European centre for particle physics near Geneva, and demonstrated the existence of the unified electroweak force.

The belief among physicists, then, is that, although the Universe is far from symmetric today, it was very different in its youth. In fact, if we could run the history of the Universe backwards like a movie in reverse, we would find that the Universe in its earliest moments was in a state of maximum possible symmetry. Every location would be like every other location, every direction like every other direction. The Universe would look the same from every conceivable point of view. It would conform to every conceivable symmetry, concrete and abstract. 'This strikes me as highly suggestive,' says Stenger. 'Why? Because the state of maximum symmetry is also the state of a completely empty void.'

The Symmetry of Nothing

Stenger amplifies what he means. Take a void, empty of matter, energy and everything else. Clearly, things look the same from every location. The void has translational symmetry. Also, things look the same in every orientation. The void has rotational symmetry. Nothing ever changes so things look the same at every time. The void exhibits time translational symmetry. In fact, the void also exhibits all the abstract symmetries of modern physics because – well, they are abstract and so do not require any matter, energy or anything.

Remarkably, then, the deep, underlying symmetries of our Universe – the symmetries which are at the root of the fundamental laws of physics – are nothing more than the symmetries of the void. 'What are we to make of this?' says Stenger.

Well, for one thing, he says, we can answer the question: Where do the laws of physics come from? 'The laws of physics are the laws of nothing,' says Stenger. And, if this is not difficult enough to stomach, there is more. If the ultimate laws of the Universe are the same as the laws of nothingness, then maybe it becomes easier to answer the arguably ultimate cosmic question: How did something come from nothing?

According to Stenger, such a transition would require no change in the laws of physics. There is no need to imagine there being no laws of physics and then the laws of physics coming into being – along with everything else – in the Big Bang. There was total continuity. The same laws of physics existed when there was nothing as they do today when there is something.

But, even though the transition from nothing to something required no change in the laws of physics, it still does not explain why the transition occurred. Why did the void not just sit there, doing nothing. Why did it change?

Stenger has an analogy. Think of water again, which has no structure, and ice, which does have structure. At the lower temperature, it is the state with structure – the state with something – that is the more stable. In this case, when the temperature drops, nothing changes into something. Stenger therefore hazards a controversial guess at why nothing turned into the something of the Universe. 'Because something is more stable than nothing!' he says.

Stenger admits he is echoing the Nobel Prize-winning physicist Frank Wilczek of the Massachusetts Institute of Technology. In 1980, writing in the magazine *Scientific American*, Wilczek pointed out that modern theories which describe how the fundamental building blocks of matter interact with each other suggest that the Universe can exist in different 'phases' analogous to the liquid and solid phases of water. In the different phases, the properties of matter are different – for instance, a particular particle might have no mass in one phase and a mass in another. And the laws of physics might be more symmetric in one phase than in another, just as the phase in which H_2O is a liquid is more symmetric than the phase in which it is ice.

Here, Wilczek makes a crucial observation. 'In these theories the most symmetrical phase of the universe turns out to be unstable,' says Wilczek.

Wilczek speculates that the Universe began in the most symmetrical state possible and that, in that state, no matter existed. A second state of the Universe was possible with less symmetry and less energy. The fact that it had less energy is crucial. In nature, things tend to seek out the lowest-energy state. For instance, a ball rolls down a slope until it reaches the lowest point, where it has the lowest 'potential energy'.

Consequently a patch of the less symmetrical phase eventually appeared and grew rapidly.

Phase changes are invariably accompanied by a release of energy. For instance, when the gas phase of water – steam – turns into the liquid phase – water, a large amount of heat is unleashed, as anyone will attest who has been scalded by the steam from a boiling kettle. According to Wilczek, the energy released when the Universe went from the symmetric state with no matter to a less symmetric state manifested itself in the creation of a storm of super-hot fundamental particles. 'This event might be identified with the Big Bang,' he says.

According to Wilczek, the fact that the Universe has no net electrical charge – that is, every particle with a negative charge is balanced by a corresponding equal but opposite positive charge – would then be guaranteed, because the Universe lacking matter had been electrically neutral. 'The answer to the ancient question "Why is there something rather than nothing?" says Wilczek, 'would then be that "nothing" is unstable.'

Or, to put it in Stenger's words: 'Something is more stable than nothing.'

Does This Really Explain Anything?

So, have Wilczek and Stenger really found a plausible answer to the ultimate cosmic question: Why is there something rather than nothing? Philosopher Stephen Law of Heythrop College at the University of London is not so sure. According to Law, the Achilles' heel of the idea that something is more stable than nothing is that it is backed up by observations of water/ice and so on – in other words, observations of the nature of its actual laws. 'The question then is: why is the Universe governed by *these* laws, rather than other laws – laws in which nothing turns out to be more stable than something?' he says. 'We are still left with an unexplained something – these laws plus the principle that nothing is less stable than something.'

Stenger believes this is to misunderstand what the laws of physics in fact are. 'The whole expression "the Universe is governed by laws" is wrong,' he says. 'The "laws" are entirely human inventions that follow from "nothing". The Universe looks just like it should if it has no laws at all.'

The laws of physics, he points out, do not follow from any unique or surprising properties of the Universe. 'Rather they arise from the very simple notion that, whatever mathematical "laws" you write down to describe measurements, your equations cannot depend on the origin or direction of the coordinate systems you define in the space of those measurements or in any abstract "space" used to describe those laws,' says Stenger. 'In other words, they cannot depend on your point of view.'

According to Stenger, except for the complexities that result from spontaneously broken symmetries, the laws of physics may be the way they are because – well – they cannot be any other way. Ultimately, there is nothing to explain. The Universe is simply structured, re-arranged nothing. We, and everything around us, are simply patterns in the void.

8

Mass Medium

Why are loaded fridges difficult to budge? Because empty space impedes them.

Photons have mass?!? I didn't even know they were Catholic.
Woody Allen

Can gravitation and inertia be identical?
This question leads directly to the general theory of relativity.
Albert Einstein, *Nature*, February 1921

What is the origin of mass? You would think physicists long ago figured out the answer to so straightforward a question. After all, they have deduced what the Universe was like in its first split-second of existence – and that must have been an incomparably harder nut to crack. Appearances, however, can be deceptive. The embarrassing truth is that, at the beginning of the twenty-first century, one of nature's best kept secrets remains the origin of mass. Remarkably, nobody knows why it is difficult to budge a loaded fridge or pick up a heavy bag of shopping.

Not only is the question of the origin of mass hard to answer, it is also open to interpretation. Just as US President Clinton said, 'I did not have sex with that woman' and qualified it with 'it all depends on what you mean by sex', physicists ask 'What is the origin of mass?' and qualify it with 'it depends what you mean by mass.' Several distinct characteristics are in fact commonly associated with mass – two familiar and everyday, and one a little more esoteric.

The Three Types of Mass

The most obvious characteristic of massive bodies is that it takes an effort to get them moving. Think how hard it is to get a broken-down car rolling. Part of the reason is of course the need to overcome the friction between the tyres and the road. But, even if there were no such force – say, the car was standing on a perfectly shiny ice rink – it would still take an effort to get the car moving.

As Galileo first realised, all bodies have a curious in-built resistance to having their motion changed in any way. Bodies which are stationary want to stay stationary; bodies which are moving want to remain moving. On Earth this is not at all obvious because frictional forces always act to slow a body down. A car whose engine cuts out, for instance, travels only a short distance before grinding to a halt. In the empty vacuum of space, however, the tendency of moving bodies to keep on moving is very apparent. Newton distilled this idea into his first law of motion, which states: 'A body remains in a state of rest or uniform motion in a straight line unless acted upon by an external force.'

This stubborn opposition of mass to any attempt to change its motion is attributed to a property called 'inertia'. Every time we try to budge a fridge and it stubbornly opposes our efforts, physicists say we are encountering the fridge's inertia. This kind of mass – 'inertial mass' – is the most familiar of all forms of mass.

In addition to inertia, however, there is another familiar characteristic which people associate with mass: 'weight'. The weight of a bag of shopping, for instance, is actually the 'force' of gravity acting on it. In everyday life we tend to use the terms weight and mass interchangeably (much to the dismay of physicists). This is possible because the force of gravity is the same everywhere on the Earth's surface so that, if one body has twice the weight of another – as shown by, say, a set of bathroom scales – we can be sure it also has twice the mass.

The fact that weight is not the same as mass but also depends on the strength of gravity was obvious to the Apollo astronauts as they bobbed about on the Moon. Although their mass was the same as on Earth, in the weaker lunar gravity their weight plummeted to a sixth of its terrestrial value.

The mass that responds to gravity is referred to by physicists as 'gravitational mass'. And there is something very peculiar about it, first noted by Galileo and others, though it took Einstein to recognise its true significance for penetrating the mystery of gravity. The peculiar thing is that the force of gravity experienced by a body goes up exactly in step with its inertial mass. In other words, a body with double the inertial mass of another, experiences twice the force of gravity; a body with ten times the inertial mass, tens times the gravity – and so on. Recall, however, that a body's resistance to being moved also goes up exactly in step with its inertial mass. Because of this, a body that is twice as hard to move as another, experiences a force of gravity twice as strong; one that is ten times as hard to move, a force ten times as great. By some weird cosmic conspiracy, the effect of inertia and the effect of gravity compensate for each other perfectly. Consequently, all bodies falling under gravity pick up speed at exactly the same rate, no matter what their mass.

Galileo is said to have demonstrated this striking property of falling bodies by dropping different masses from the top of the Leaning Tower of Pisa and seeing them hit the ground simultaneously. However, it was shown beyond a doubt in 1972 when Dave Scott, commander of the Apollo 15 spacecraft, dropped a hammer and a feather together in the air-resistance-free environment of the Moon. Two simultaneous puffs of dust on the fuzzy black-and-white TV pictures were indisputable evidence that the hammer and the feather had struck the lunar surface at the same time.

Nobody knows why 'gravitational mass' – which determines the gravitational force a body experiences – is the same as the inertial mass – which determines a body's resistance to being moved. Nevertheless, in 1915, Einstein was able to use this 'equivalence' of gravitational mass and inertial mass as the cornerstone of his theory of gravity, the general theory of relativity.*

But, in addition to gravitational mass and inertial mass, there is a third, less familiar, type of mass. This is the one physicists believe they have made the most progress in understanding. It stems from a more esoteric characteristic of matter. As well as responding to gravity and resisting attempts to change its motion, mass also behaves as a super-concentrated 'knot' of 'energy'.

Energy is actually a slippery concept to define but it comes in many forms – for instance, heat energy, sound energy and energy of motion. If you doubt that energy is associated with motion, step into the path of a speeding bicycle – or, better still, just imagine it! As pointed out before, one of the central characteristics of energy – first appreciated by nineteenth-century scientists and enshrined in the law of conservation of energy – is that energy can never be created or destroyed, merely changed from one form into another.

---

* Einstein's genius was to realise there is actually no 'force' of gravity. Falling bodies simply move under their own inertia. The twist is that they are not moving in straight lines at constant speed, as Newton thought, but along the shortest possible paths through the warped landscape of four-dimensional space-time – something beyond our powers to visualise (see Chapter 7, 'Patterns in the Void').

In 1905, Einstein surprised pretty much everyone by identifying an entirely new form of energy – the energy associated with mass. Of all the forms of energy, mass-energy is by far the most concentrated. A single gram of matter contains as much energy as is liberated by the detonation of about 10 tonnes of dynamite. (The precise amount of energy, E, locked inside a chunk of matter of mass, m, is given by the most famous equation in the whole of science: $E = mc^2$, where c is the speed of light.) This fact would be of little consequence if the energy in matter was locked away irretrievably. However, it is not. Because one form of energy can be transformed into another, mass-energy can be converted into other forms of energy, ultimately heat energy. The awful reality of this was demonstrated in August 1945 with the dropping of atomic bombs on the Japanese cities of Hiroshima and Nagasaki.[*]

One of the key questions in physics is: Why does matter have the mass-energy it does? The question can be made more precise by relating it to the fundamental building blocks of all matter. These appear to be six particles called quarks and six particles called leptons, with only two types of quark and two types of lepton involved in the construction of all normal matter, including you and me.[†] The crucial question physicists want to answer is therefore: Why do the fundamental particles have the masses they do? For instance, why does a top-quark have about a million times the mass-energy of an electron?

---

[*] If mass-energy can be converted into other forms of energy, it follows that other forms of energy can be transformed into mass-energy. This is routine at giant particle accelerators like the one at CERN. Super-fast beams of subatomic particles are smashed together and their energy of motion converted into the mass-energy of new particles, which are conjured into existence like microscopic rabbits out of hats.
[†] Actually, physicists suspect that, in addition to the ordinary matter in the Universe, there is about six to seven times as much mass in invisible, or 'dark', matter. A strong possibility is that it consists of hitherto unknown fundamental particles.

The Origin of Mass-Energy

The fundamental particles of matter are believed to be 'glued' together by just four fundamental forces. These are the electromagnetic force, which binds together the atoms in our bodies; the weak nuclear force and the strong nuclear force, which hold sway in the ultra-tiny domain of the atomic 'nucleus'; and the gravitational force, which governs the large-scale Universe of planets and stars and galaxies. All these forces are believed to arise from the exchange of 'force-carrier' particles, which shuttle back and forth between the fundamental particles like microscopic tennis balls between tennis players. In the case of the electromagnetic force, for instance, the force-carrier is the photon, the humble particle of light; in the case of the strong nuclear force it is the 'gluon', which comes not in one but eight different types.

Having four different fundamental forces, transmitted by a legion of force-carrying particles, may appear a bit on the complicated side. And this is the consensus view of physicists. They are convinced that things were very much simpler once upon a time. What encourages them in this belief is nature's peculiar obsession with symmetry. All the fundamental laws of nature seem to be mere manifestations of deep underlying symmetries of the world – properties which remain the same when our viewpoint is changed, either in some concrete or in some abstract way.* Admittedly, in today's world, some of nature's symmetries are flawed, or broken. Nevertheless, physicists believe that, in the blistering hot fireball of the Big Bang, symmetry reigned supreme.

The evidence for this belief rests on the observation that systems become simpler – which is synonymous with more symmetric – at

---

* See Chapter 7, 'Patterns in the Void'.

higher temperatures. Ice, for instance, is not the same throughout – often it contains bubbles and flaws and fissures. However, if ice is heated until it becomes water, the non-uniformities vanish. It looks the same at every place, which is the same as saying it is more symmetric.

Similarly, physicists believe that, when the Universe was a lot hotter – as it was in the first moments of the Big Bang – it was a lot simpler, a lot more symmetric. Instead of the four fundamental forces, there was a single 'superforce'. The fundamental forces we observe in today's Universe turn out to be mere facets of this 'unified' force.

In fact, this belief is more than a qualitative, hand-waving one. As pointed out before, in the mid-1980s, at CERN, the European centre for particle research near Geneva, physicists slammed together subatomic particles with such violence that they were able to recreate for a split-second the ultra-high temperatures which existed in the fireball of the Big Bang. Sure enough, they observed the electro-magnetic and weak forces merge together into a single, electroweak, force. Theorists have also devised 'Grand unified theories', or GUTs, which predict that, at the far higher temperatures that existed at even earlier times than the Big Bang, all the non-gravitational forces of nature – not just the electromagnetic and weak forces – were fused together into a single force.

All this is well and good. However, there is a rather serious snag with GUTs and their prediction of the unification of the forces. The unific-ation could have happened only if the fundamental particles had no mass!

Clearly, today's fundamental particles are not massless. Consequently, a vital jigsaw piece is missing from our picture of reality. Specifically, there must exist a mechanism by which nature bestows mass on the massless particles. Such a mechanism was in fact fleshed out in the mid-1960s by the Scots physicist Peter Higgs.

The Higgs mechanism employs the 'Higgs field', a subtle and

previously unsuspected feature of the vacuum.* Physicists often visualise it as a kind of invisible cosmic treacle which fills all of space and sticks to particles. Since the Higgs field contains energy – which has a mass equivalent – the more treacle a particle accrues, the more massive it is. 'The picture of particles accumulating some treacle is crude – but not totally misleading,' says Nobel Prize-winner Frank Wilczek of the Massachusetts Institute of Technology.†

The mass the Higgs field bestows on a particle is called its 'rest mass'. This is the mass it possesses when it is standing still and it is intrinsic to the particle. It is important to recognise this because a particle may also possess a mass by virtue of being in motion. This is apparent from Einstein's $E = mc^2$ formula, which can equally well be read from left to right as right to left. Not only is mass a form of energy – mass-energy – but energy has an equivalent mass. Since one form of energy is energy of motion, a particle has a greater mass when it is moving than when it is not.‡

Fields like the Higgs field, which fill all of space like a vast sea, are actually considered by physicists to be the most fundamental things in nature – even more fundamental than particles. In the modern view, the particles are nothing more than localised 'excitations' – vortices if you

* A field is simply an entity which fills all of space, assigning to each point a unique property. In the case of the gravitational field, for instance, the property is the direction and strength of the force of gravity.

† Wilczek shared the 2004 Nobel Prize for Physics with David Gross and David Politzer for the discovery of 'asymptotic freedom'. This is the idea that quarks behave like free particles at ultra-high energies, which is synonymous with the particles being very close together. If the quarks are pulled apart, however, they feel a force so tremendous that it is impossible to separate them completely. Wilczek's discovery, made in 1973, led to Quantum ChromoDynamics (QCD), the theory of the strong nuclear, or 'colour', force that binds together quarks. It was an important plank in the Standard Model that describes the three non-gravitational forces of nature.

‡ Interestingly, the photon – the particle of light – has zero rest mass. This is because it cannot stand still. It is born travelling at the speed of light and all its energy is energy of motion.

like – in that sea. The vortex of the Higgs field is not surprisingly called the Higgs particle. In fact, there may be several different types of Higgs particle, the number depends on precisely how nature has chosen to implement the Higgs mechanism.

Particle physicists are desperate to find the Higgs particle, which is widely regarded as a missing piece of their 'Standard Model' of fundamental particles and forces. Currently, all their hopes are pinned on CERN's 'Large Hadron Collider' which, when completed in 2008, will be the world's most energetic particle accelerator. If the LHC finds the Higgs, the relief among physicists will be palpable. Almost certainly they will declare to every newspaper and TV station that they have found the elusive origin of mass. However, according to Wilczek, this will be an exaggeration of the truth. 'You have to be very clear about exactly what the Higgs mechanism does and does not explain,' he says.

The Higgs mechanism does not in any sense 'explain' the actual values of the masses of the fundamental particles. For instance, it does not explain why the top-quark has roughly a million times the mass of the electron. In the Higgs theory, the value of the mass of a particular particle depends on how well the Higgs 'treacle' sticks to it. The stickiness is encapsulated in a number called a 'coupling constant', which is different for each particle and, worse still, must be inserted into the theory by hand simply to make the masses come out right.

Far from shedding light on the origin of mass, the Higgs mechanism appears merely to substitute one mystery for another. Instead of having to explain the different masses of the fundamental particles, physicists must explain their different coupling constants. 'I guess whether or not you call the Higgs mechanism an explanation of mass is a matter of taste,' says Wilczek. 'I would be inclined to say no, since it doesn't simplify the description of mass, nor suggest testable new properties of mass.'

Even if the Higgs mechanism is accepted as an explanation of mass, it has another, rather embarrassing, shortcoming. It can account for hardly any of the mass of ordinary matter – that is, of you and me!

Most of the mass of ordinary matter is tied up in particles called protons and neutrons, which are found in tight clumps at the heart of atoms.[*] The protons and neutrons are, in turn, composed of up-and-down-quarks. The Higgs mechanism can certainly account for the masses of these quarks. However, the quark masses are actually very small indeed. In fact, they contribute only a tiny fraction of the mass of protons and neutrons. So, even with the Higgs, the lion's share of their mass is still unaccounted for. It must arise in some totally different way.

In fact, the missing mass arises from the gluons, which transmit nature's strong, or 'colour', force, and so glue together the quarks inside protons and neutrons.[†] The bizarre thing is that gluons actually have no mass whatsoever! Nevertheless, the colour field possesses mass by virtue of the fact it contains energy, much as the magnetic field of a magnet contains energy – witness its ability to draw in metal filings. The mass of the magnetic field of even the strongest magnet is far too small to be measurable. However, the colour force is enormously stronger than any conceivable magnetic force. Consequently, the mass of the colour field is

[*] In an atom, a tight ball of protons and neutrons – in the case of the lightest atom, hydrogen, a single proton – is bound together into an ultra-compact 'nucleus'. The nucleus accounts for 99.9 per cent of the mass of an atom, with the remainder accounted for by electrons which flit about the nucleus in a diffuse 'cloud' about 100,000 times bigger than the nucleus.

[†] The force is called the colour force because particles react to it according to their 'colour'. This is not a real colour. It is analogous to the electric charge of the electromagnetic force. But, whereas electric charge comes in two types – negative and positive – the 'charge' of the strong force comes in three types. This is one reason why physicists have chosen the primary colours red, green and blue. The other – cleverer – reason is that, with this choice, all the stable composite particles are colourless. For instance, a proton consists of three quarks, one in each primary colour, and the three primary colours cancel each other out.

substantial. In fact, it can perfectly account for the missing mass of ordinary matter.

Bizarrely, then, most of the mass of you and me does come not from the fundamental building blocks of our bodies. Instead, it comes from the glue that cements those building blocks together.

The inability of the Higgs mechanism to explain much of the mass of ordinary matter – or even predict the precise values of the masses of the fundamental particles – might be expected to dampen the enthusiasm of physicists. Not a bit of it. 'A lot of hype is perpetrated about the Higgs mechanism and what it actually explains,' admits Wilczek.

Without doubt, the Higgs mechanism provides a vivid and intuitive picture of how ordinary matter acquires a small portion of its mass by accumulating the cosmic treacle of the Higgs field. However, the kind of mass being talked about is, strictly speaking, the most esoteric type of all – a measure of the energy content of matter. Think of it as microscopic book-keeping. If a particle at rest disintegrates, or 'decays', into other particles, the total energy of these particles must always be equal to the mass-energy of the original particle.[*]

But the energy content of mass is, of course, only one of its characteristics. Mass also opposes attempts to change its motion and reacts to gravity. Where do inertial mass and gravitational mass come from? Like most physicists, Wilczek thinks these types of mass come along, part and parcel, with mass-energy. In other words, the Higgs mechanism explains not just the energy content of mass but all aspects of mass. But others disagree. 'There is nothing in the Higgs theory that explicitly says mass-energy should doggedly oppose all attempts to change its motion or that

---

[*] In fact, such microscopic transformations are subject not only to the law of conservation of energy but to other edicts of microscopic book-keeping – such as law of conservation of momentum, electric charge, and so on. All of these must be obeyed simultaneously.

it should respond in any shape or form to gravity,' says Bernard Haisch of the Calphysics Institute in Scotts Valley, California.

Haisch is perfectly prepared to believe that the rest mass of a particle – its mass-energy – is 'explained' by the Higgs mechanism and that the rest mass is intrinsic to the particle. However, he believes that the inertial mass and gravitational mass of a particle are not explained by the Higgs mechanism and are not intrinsic. If they are not intrinsic then there is only one other option. They must be 'extrinsic'. 'In other words, they must somehow arise from the interaction between a particle and the environment through which it moves,' says Haisch. 'That environment can only be the "quantum vacuum".'

The Quantum Vacuum

The quantum vacuum is an unavoidable consequence of two things, the first of which is the existence of fields of force. As pointed out, physicists view fundamental reality as a vast sea of such fields. In their picture, known as 'quantum field theory', the fundamental particles are mere localised humps, or knots, in the underlying fields.

The best understood of all the fields, and the one with the greatest bearing on the everyday world because it glues together the atoms in our bodies – not to mention all other normal matter – is the electro-magnetic field. The electromagnetic field can undulate in an infinite number of different ways, each oscillation 'mode' corresponding to a wave with a different wavelength.* Think of the waves at sea, which can range all the way from huge, rolling waves down to tiny ripples.

* The wavelength is the distance over which a wave goes through a complete oscillation cycle - twice the peak-to-peak distance.

Naively, the vacuum of empty space would be expected to contain no electromagnetic waves whatsoever. And this would be true but for the small matter of the Heisenberg uncertainty principle. According to the principle, every conceivable oscillation of the electromagnetic field must contain at least a minimum amount of energy.[†] This seemingly innocuous rule has dramatic and profound implications for the vacuum because it means that each of the infinite number of possible oscillation modes of the electromagnetic field must be jittering with the minimum energy dictated by the uncertainty principle. In other words, the existence of each mode is not simply a possibility, it is a certainty. Far from being empty, the 'quantum vacuum' is a fantastically choppy sea of fluctuating fields.[‡]

These quivering fields, known as 'quantum', or 'zero-point', fluctuations, can manifest themselves in a truly remarkable way. Recall that fundamental particles are nothing more than localised hummocks in

---

Actually, each oscillation mode corresponds to a wave with a different wavelength and 'polarisation'. A wave's polarisation is related to the 'plane' in which it vibrates.

[†] The Heisenberg uncertainly principle actually states that, for any microscopic event, there is a minimum value of particular quantity – the duration of the event multiplied by the energy of the event. An oscillation has a characteristic time associated with it – the duration of a single oscillation – so the Heisenberg uncertainty principle dictates it must also have a certain minimum energy associated with it. If this sounds gobbledygook, think again of that teenager who borrows his dad's car and gets it back in the garage before his dad gets up in the morning so that, as far as his dad is concerned, the car was never borrowed. This is the way it is with energy. It can be 'borrowed' just as long as it is given back before nature notices it's missing. The smaller the quantity of energy the longer it takes nature to notice, and vice versa. To put it bluntly, the Heisenberg uncertainty principle is merely a measure of nature's blindness.

[‡] It would seem that an infinite number of oscillation modes, each with non-zero energy, adds up to an infinite amount of energy in every chunk of the vacuum. However, physicists believe there is a 'cut-off' at a particular super-high energy

nature's underlying fields. Consequently, the choppy sea of the vacuum is continually conjuring particles into existence like microscopic rabbits out of hats. Known as 'virtual particles', they have only a fleeting existence, popping into existence for far less than the blink of an eye before popping back out again.

Haisch began thinking about the intriguing possibility that the quantum vacuum might have something to do with inertial mass in February 1991. The trigger was a talk he attended by Alfonso Rueda of California State University in Long Beach. It was about 'stochastic electrodynamics'.

According to the idea, first devised in the 1960s, the quantum vacuum is absolutely central to the creation of the world. Ultimately, all the bizarre 'quantum' behaviour of microscopic particles can be traced back to the relentless buffeting they receive from the ceaselessly churning quantum vacuum.

Sitting in the audience of Rueda's talk, Haisch had been pretty much letting it all waft over him when something Rueda said suddenly made him sit up and pay attention. Rueda mentioned an esoteric discovery made independently by two physicists in the 1970s. Paul Davies and Bill Unruh had been exploring Stephen Hawking's remarkable idea that black holes are not completely black. According to Hawking, the immense gravity close to a black hole distorts the quantum vacuum in such a way that the fleeting virtual particles can become real, fountaining out of the vacuum as permanent 'Hawking radiation'.

Understanding exactly how Hawking radiation arises requires knowing that, when virtual particles pop out of the vacuum, they pop out in pairs – typically an electron–positron pair. The positron is the

density known as the 'Planck energy'. Modes with energy densities greater than this are simply not permitted to exist by nature.

'antiparticle' of the electron. Every subatomic particle has an associated antiparticle with opposite properties such as electrical charge. A particle and its antiparticle are always born together. Furthermore, when a particle and its antiparticle meet, they self-destruct, or 'annihilate'. Such creations and annihilations are the stuff of the quantum vacuum. All over space electron-positron pairs are continually winking into existence, lingering in the world for the merest of instants, then undergoing mutual annihilation and winking out again.

Close to a black hole, however, something else can happen. During the fleeting life of an electron-positron pair, one of the particles can find itself dragged through the black hole's 'event horizon' – the point of no return for in-falling matter. Since the particle outside the hole now has no partner with which to annihilate, it has no means of popping back out of existence. It has been elevated from the status of a transitory virtual particle to a real particle with a permanent existence.

All around the horizon of a black hole virtual particles from the quantum vacuum are continually being boosted to reality in this way, flying away from the hole as Hawking radiation. Of course, ultimately something must pay for their mass-energy and that something is the gravitational field of the black hole, which gradually weakens as it loses an equivalent amount of energy.*

What Davies and Unruh were interested in was exactly what the Hawking radiation would look like. They concluded that an observer looking at the black hole would see radiation exactly like that which emerges from a hot furnace. In the case of a furnace, the radiation is known as 'thermal' radiation and has a mix of colours determined solely

---

* Hawking radiation is extremely weak, and pretty much undetectable, for a stellar-mass black hole. However, microscopic black holes would shine brightly with Hawking radiation. These might have been made when matter was squeezed in the violence of the Big Bang.

by the temperature of the furnace. In the case of the black hole, the mix of colours is determined by the black hole's gravity. In some weird sense, then, a black hole's gravity gives it a 'temperature'. Black holes are hot!*

From this unexpected discovery, Davies and Unruh made an intriguing extrapolation. Einstein had realised that gravity is indistinguishable from constant acceleration, at least in any small enough region of space. This he called 'the happiest thought of my life' and made it a cornerstone of his theory of gravity.† The equivalence of gravity and acceleration enabled Davies and Unruh to extend their result about Hawking radiation. Just as someone near a black hole would see heat radiation coming from the hole's vicinity with a temperature dependent on its gravity, someone accelerating through space – that is, through the quantum vacuum – would see heat radiation coming from in front of them with a temperature dependent on their acceleration.

The virtual particles popping in and out of existence in the quantum vacuum actually have a remarkable property. If an observer flies through the vacuum at constant speed – and it does not matter what that speed is, as long as it is constant – the view they see of virtual particles popping in and out of existence is the same behind them as in front of them. This means that the quantum vacuum is completely compatible with Einstein's special theory of relativity, which recognises that all observers travelling through space at uniform speed see the world in exactly the same way.

Because the vacuum looks the same to an observer flying through it

---

* The discovery that black holes have a temperature means that in black holes three of the great theories of physics link up – 'thermodynamics', the theory of heat; quantum theory, the theory of the microscopic world; and general relativity, Einstein's theory of gravity. This is believed to be of profound significance but nobody yet knows what that significance is.

† Einstein elevated the indistinguishability of gravity and acceleration to a grand principle of physics, the 'principle of equivalence'. It recognises that gravity is not even a real force. It exists only because we are accelerating – only we don't realise it.

at constant speed, it effectively behaves as if it is not there. Just like the air in which we live, it has no discernible effect on us. However, Davies and Unruh's discovery that someone who accelerates through the quantum vacuum will find themselves bathed in heat shows that accelerated motion is fundamentally different from motion at constant speed. From the point of view of an accelerated observer, the vacuum is transformed into a real, detectable thing, capable of affecting them.

To Haisch, listening to Rueda's talk, this prompted an intriguing thought. 'If an accelerated body sees heat coming at it from in front, that heat might apply a force which slows the body,' says Haisch. 'I'm an astrophysicist, you see. I'm used to the idea that heat radiation – for instance, sunlight – can exert a force on bodies such as the tiny dust particles that make up the tail of a comet.'

After the talk, Haisch told Rueda his idea and Rueda said he would do some calculations. For a few months nothing happened. Then, one morning, Haisch arrived at his office at Lockheed Martin's Solar and Astrophysics Laboratory in Palo Alto to find his answering machine light flashing. 'Alfonso had left a message at three a.m.!' says Haisch. 'He was so tremendously excited by the result of a mammoth calculation he had been doing that he couldn't wait to tell me. "I think I can explain Newton's second law of motion!" he said.'

Newton's second law, postulated in 1687, is conventionally written as $F = ma$, where F is the force experienced by a body of mass, m, and a is the acceleration that results. The law is in fact no more than a definition of inertial mass, which is defined as the ratio of the force applied to a body to the acceleration produced.

Rueda had examined in detail the electric and magnetic components of the electromagnetic radiation experienced by a body as it accelerates through the vacuum. A magnetic field has long been known to exert a force on a moving electric charge. In fact, this is the basis of the electric motor. When an electric current – a flow of charged electrons – passes through a coil of wire placed in a magnetic field, the coil rotates on its spindle. Rueda discovered that there would be a similar force between the magnetic field experienced by an accelerating body and the moving electric charges in the atoms of the body. 'And when he calculated the force he found it was exactly as required by Newton's second law,' says Haisch. 'A retarding force which depended on the body's acceleration. After three centuries, someone had at last explained inertia.'

According to Rueda, inertial mass is not intrinsic to a body at all. It is extrinsic, bestowed on a body from outside. Specifically, it arises from the interaction between the basic building blocks of matter and the great roiling ferment of virtual particles that make up the quantum vacuum.[*]

Haisch says he is not surprised by the idea that inertial mass is not intrinsic to a material body – that it is not a fundamental thing. He points to the fact that inertial mass is impossible to measure directly. Instead, people infer it. They take two measurable quantities – the force applied to a body and the acceleration it produces – and deduce the mass of the body from the ratio. The lot of inertial mass, Haisch believes, is to go the same way as space and time. In the wake of Einstein's special theory of relativity, both ceded their fundamental status to the speed of light. 'Inertial mass is not a fundamental thing,' says Haisch. 'The really fundamental thing turns out to be the quantum vacuum.'

If mass is not a fundamental thing, it may explain why it appears to come in so many different kinds – inertial mass, gravitational mass and rest mass, the mass associated with energy. 'It simply reveals a different face depending on how it is measured,' says Haisch.

So, what of the Higgs mechanism? Haisch sees no incompatibility

---

[*] See *Physical Review A*, vol. 49, p. 678.

between this and the electromagnetic interaction between a particle and the vacuum. 'The Higgs mechanism explains the rest mass of subatomic particles while the vacuum interaction explains their inertial mass,' he says.

Rueda agrees. 'The Higgs field deposits energy, and hence rest mass, around structures we call elementary particles,' he says. 'The claim that this accumulated energy behaves in some way that gives such elementary particles the property of inertia is a mere hope. You need something else for that – and we think we've found it.'

As mentioned earlier, the force carrier of the electromagnetic field is the photon. At a microscopic level, therefore, the interaction between the constituent particles of matter and the quantum vacuum involves photons being exchanged between the virtual particles of the vacuum and the quarks and electrons in matter.

An electron is considered by physicists to be a truly fundamental and indivisible particle – a point-like concentration of electric charge. However, in order to obtain his $F = ma$ result, Rueda had to assume that such an electron jitters back and forth within a characteristic volume of space.[*] This may seem a bit of an arbitrary – not to say peculiar – assumption. However, it revives an old idea proposed by Louis de Broglie and Erwin Schrödinger, two of the pioneers of quantum theory.

De Broglie and Schrödinger were puzzled that, in experiments in which photons rebound, or 'scatter' off electrons, the electrons behave exactly as if they have a particular size called the 'Compton wavelength'. To make sense of this, the two physicists proposed that an electron is in fact a point-like charge which jitters about randomly within a sphere of diameter the Compton wavelength. They called this

[*] See *Physics Letters A*, vol. 240, p. 115.

trembling motion 'zitterbewegung'. 'Alfonso and I believe De Broglie and Schrödinger were onto something with their zitterbewegung,' says Haisch. 'Their mistake, however, was in thinking that the motion was intrinsic to an electron. In fact, it is extrinsic – due to the random battering the point-charge receives from the jittery vacuum. In effect, this smears out the electron, making it appear as big as the Compton wavelength.'

According to Haisch, it is always possible that this jittering motion could explain more than inertial mass. 'A massless particle may pick up energy from the zitterbewegung, hence acquiring what we think of as rest mass,' he says. 'It would be a neat, tidy package. It might be possible to dispense with the Higgs mechanism altogether. It strikes me as far more elegant than an undetected Higgs field.'

Piling speculation on speculation, Haisch and Rueda suspect that the interaction that produces inertia occurs preferentially at a special, 'resonant', frequency. This is a frequency at which energy is most efficiently transferred from one body to another. Think of someone pushing a swing. Everyone knows there is a particular frequency – perhaps one once every ten seconds – at which the energy in the push is most effectively transferred to the child on the swing, making it go higher and higher. This frequency of once every ten seconds is an everyday example of a resonant frequency. Well, when Haisch and Rueda speculate that the interaction that produces inertia occurs preferentially at a resonant frequency, they speculate further that this resonant frequency is the zitterbewegung, or Compton, frequency. 'If we knew what caused this resonance, we would probably be able to explain the ratio of the various quarks rest masses to the electron rest mass,' says Haisch.

If, as Haisch and Rueda believe, inertial mass is a consequence of an electromagnetic interaction with the vacuum, this still cannot explain the small mass claimed for a particle such as the 'neutrino'.* This is because it interacts via the weak nuclear force and not the electromagnetic force. 'The origin of neutrino mass must be in its interaction not with the electromagnetic zero-point fields of the vacuum but with the zero-point weak fields,' says Haisch.

Gravitational Mass

If inertial mass does indeed have its origin in the interaction between matter and the quantum vacuum, what of gravitational mass? Well, inertial and gravitational mass are of exactly the same magnitude, an observation which is a cornerstone of general relativity. This equivalence can logically have only a limited number of possible explanations.

One is that inertial mass has a gravitational origin. This was the hope of the nineteenth-century Austrian philosopher Ernst Mach. He postulated that inertia of a body was the result of the combined gravity of all the objects in the Universe. The reason there is resistance when you try to stop a moving body or start a stationary body, Mach maintained, is because the stars and galaxies of the Universe are pulling against you!

Mach's idea appealed enormously to Einstein. He hoped that it would emerge as a natural consequence of his own theory of gravity. However, he was to be disappointed. Like everyone else, Einstein was reduced to assuming, without any understanding or proof, that matter has inertia.

---

* A neutrino is a neutral subatomic particle with a very small mass that travels very close to the speed of light and hardly ever interacts with matter. Something like 100 million million neutrinos from the Sun pass through every square centimetre of your body every second.

Nowadays, Mach's idea has fallen out of favour, principally because it requires the Universe to react instantaneously to the acceleration of a body on Earth. However, we are pretty sure that the cosmic speed limit is set by the speed of light and that no influence, not even gravity, can act without any time delay.

A second logical possibility for the equivalence of inertial and gravitational mass is that gravitational mass has an inertial origin. In fact, this is what Einstein showed in general relativity. There is in fact no 'force' of gravity. Bodies actually move under their own inertia along straight lines. The straight lines, or 'geodesics', are actually in a higher, four-dimensional, space-time and so appear to us as curves. However, even though general relativity shows that gravitational mass has an inertial origin, the theory still leaves unanswered the question: What is the origin of inertial mass? 'Trying to coax inertia out of gravity or gravity out of inertia, you wind up with an inevitable circularity,' says Haisch.

The final logical explanation for the equivalence of inertial and gravitational mass is that they share a common origin. And this is what Haisch and Rueda think: both kinds of mass, they claim, arise from interactions of the electric charges of matter with the quantum vacuum. But, whereas Haisch and Rueda's idea of the origin of inertial mass is well developed, their idea of the origin of gravitational mass is far more speculative.

Basically, the two physicists believe that charges in a chunk of matter distort, or 'polarise', the quantum vacuum in their immediate vicinity. In other words, they attract virtual particles with opposite electrical charges and repel virtual particles with similar electrical charges. This distortion of the vacuum in turn interacts with the charges in another chunk of matter. By this roundabout means, a force of attraction arises between the two chunks. 'The mechanism is so tortuous it might explain why gravity is so much weaker than the other fundamental forces of nature,' says Haisch. 'One mass does not pull directly on another mass but only through the intermediary of the quantum vacuum.'

Haisch and Rueda's description may appear puzzling if you know anything about Einstein's theory of gravity. After all, general relativity 'explains' gravity perfectly in terms of the warpage of higher-dimensional space-time by matter. At first glance, this 'geometrical' picture does not appear to be at all compatible with the picture of Haisch and Rueda.

However, Haisch points out that the warpage of space described by Einstein's theory is actually not directly measurable. Instead, astronomers infer it from the bending of the paths of light rays passing through space. If the light from a distant star passes close to the Sun on its way to the Earth, for instance, its path is bent by the warped space close to the Sun. 'If matter distorts, or "polarises", the quantum vacuum, this changes its ability to bend light, or its "refractive index",' says Haisch. 'The vacuum then bends the path of light just like a piece of glass does.'

Haisch conjectures that the change of refractive index of the vacuum caused by the presence of matter has exactly the same effect on the paths of light rays as the warpage of space which in Einstein's theory is caused by the presence of matter. In this way, all the mathematics of general relativity remains intact since space-time, though unwarped, looks exactly as if it is warped!* 'I strongly suspect that the vacuum-inertia theory can be made consistent with general relativity and the warping of space-time,' says Rueda. 'But it is still too early to be certain.'

In their latest work, Rueda and Haisch even explain why inertial mass

---

* Changes in the refractive index of the vacuum resulting in a pseudo-curved space-time goes by the name of 'polarisable vacuum'. Those responsible for work on this idea are principally Hal Puthoff and Robert Dicke.

and gravitational mass are the same. And it turns out to be remarkably straightforward. If you accelerate through the quantum vacuum, the vacuum resists your motion, which is why you have inertia. However, if you are held fixed in a gravitational field, it is the quantum vacuum that accelerates past you. 'But this immediately shows that the "mass" associated with inertia and the "mass" associated with weight must be equal because the two situations are the same,' says Haisch. 'Accelerating through the quantum vacuum or having the quantum vacuum accelerate past you are the same process. Hence Einstein's principle of equivalence is neatly explained.'*

Perhaps the most mind-blowing consequence of gravitational and inertial mass owing their existence to the vacuum is the possibility of modifying both through modifying the vacuum. If a way could be found to change the vacuum in the right way, it might be possible to nullify mass, making an inertia-less drive that could accelerate a spaceship from a standstill to the speed of light – the cosmic speed limit – in the blink of an eye!

---

* 'Gravity and the Quantum Vacuum Inertia Hypothesis' by Alfonso Rueda and Bernard Haisch (*Annalen der Physik*, vol. 14, no. 8, p. 479, 2005).

Part Three
LIFE AND THE UNIVERSE

9

An Alien at My Table

Will we ever find ETs out in the Universe?
No – but we might find them inside
a computer!

Where is everybody?
Enrico Fermi, 1950

The fact that we have not yet found the slightest evidence for life – much less
intelligence – beyond this Earth does not surprise or disappoint me in the
least. Our technology must still be laughably primitive, we may be like jungle
savages listening for the throbbing of tom-toms while the ether around them
carries more words per second than they could utter in a lifetime.
Odyssey: the Unauthorised Biography of Arthur C. Clarke, Neil McAleer, 1992

Extraterrestrial intelligences are out there. But the people involved in the
search for them are going about it entirely the wrong way. Even if they
were going about it the right way, the task would be nigh on impossible.
And, even if it was not impossible, and 'contact' was miraculously made,
there would be nothing much we could learn from extraterrestrials that
we could not discover ourselves simply by playing about on a desktop
computer!

This is the controversial opinion of Stephen Wolfram, evangelist of the
idea discussed earlier that nature has created all the bewildering variety
of the Universe simply by repeatedly applying a simple complexity-

generating computer program.* Wolfram is not saying this just to be controversial. Once upon a time, he was an enthusiast for SETI, the search for extraterrestrial intelligence. 'It's just that the work I've done over the past twenty years has taught me that SETI as currently practised is not the most sensible thing to do,' he says.

Of course, it is not at all obvious that the sensible thing to do is to abandon any systematic scanning of the heavens for signs of extra-terrestrials and instead look inside a computer. In fact, the suggestion seems utterly mad. But Wolfram has his reasons. And the first step in appreciating those reasons, he says, is to understand what is wrong with the current strategy for finding extraterrestrial intelligence.

An Extraterrestrial Signal

Take the search for alien signals. Electromagnetic waves are continually raining down on the Earth from the Universe. These span a vast range, or 'band', of frequencies.† For instance, there are radio waves – the most sluggishly oscillating waves – and gamma rays – the most rapidly oscillating waves. The visible, or 'optical', light picked up by the human eye is pretty much in the middle of this electromagnetic 'spectrum'.

SETI scientists focus their attention on the radio and optical bands for the simple reason that both types of wave can travel across large swathes of the Galaxy without being absorbed and that both can easily penetrate the Earth's atmosphere. In effect, SETI researchers look for the interstellar equivalent of a radio station. This is an electromagnetic wave

* See Chapter 2, 'Cosmic Computer'.
† The frequency of a wave is simply how fast it oscillates up and down. Normally, this is measured in Hertz (Hz), where 1 Hz is 1 oscillation per second.

which spans a narrow range of frequencies and which carries information – the information is impressed on the 'carrier wave' by continuously varying, or 'modulating', its frequency or amplitude.*

In short, what SETI scientists do is sift through the chaotic babble of radio 'static' coming from space in the hope of finding an interstellar radio station – a signal with some kind of regularity, some kind of pattern. 'But a patterned signal is an inefficient signal,' says Wolfram.

Take, for instance, a patterned signal which consists of a 1 followed by a 0 a billion times over – in other words, 1010101010 . . . It would take ages to send. It would be terribly inefficient. However, the pattern can be exploited to compress the signal down to something far more efficient that can be transmitted far more quickly: 'Repeat 10 a billion times'.

The trend towards more efficient communications is evident today in terrestrial communications. The need to cram as much information as possible into a signal has led to communication technologies in which information is sent using a vast number of frequencies simultaneously. Take CDMA – Code Division Multiple Access – the cell phone standard used in North America. Here, the carrier, rather than being a periodically undulating wave, is a pseudo-random sequence of binary digits – something a lot more complicated. 'The key point is that the signal being transmitted has very little pattern – it looks almost random,' says Wolfram.

There is a lesson here for SETI, says Wolfram. Because a patterned signal is an inefficient signal, it is very unlikely to be used by an advanced technology. Instead, Wolfram maintains, the transmissions of an advanced civilisation will be as far from the patterned signals that people are searching for as it is possible to imagine. In fact, the transmissions will look pretty much like the random radio 'static' which comes from astronomical

---

* The amplitude of a wave is its maximum excursion from its average level. Think of it as the wave's height.

objects such as stars and interstellar clouds of gas. 'Distinguishing a signal from an extraterrestrial intelligence from the natural background of cosmic signals is going to be extremely difficult,' says Wolfram.

This may explain why SETI, after more than four decades of effort, has drawn a disappointing blank. As the Italian-American physicist, Enrico Fermi, asked: 'Where is everybody?' If Wolfram is right, extra-terrestrials are out there but they are a lot more difficult to spot than anyone ever imagined.

In fact, Wolfram can see only one way to distinguish a signal from an extraterrestrial intelligence from the natural background – and that is to use a tedious process of elimination. Consider the radio emissions our radio telescopes pick up from our Galaxy as a whole. This consists of the sum total of the radio waves emitted by all the objects in the Milky Way – for instance, the radio static from the nearest star, Alpha Centauri; the radio static from the Crab Nebula; the radio static from the clutch of newborn stars in the Pleiades; and so on. According to Wolfram, we have no choice but to compare the sum of this static with the static we pick up from the Galaxy as a whole. 'Only if there is some radio static left over can we say we *may* have picked up an intelligent signal,' says Wolfram.

The 'may' here is the operative word. The leftover radio emission might actually have a natural cause but one which we have not yet identified. 'Even if we were to pick up a cosmic radio signal containing the digits of pi, it is always possible that a natural process could be generating it,' says Wolfram.[*] According to Wolfram, we may never be able to exclude the possibility that a signal we have detected is natural. 'It is very unlikely that the question "Is this an intelligent signal?" will have a simple yes/no answer,' he says.

[*] This alludes to the fact that the number pi, though it appears complex, its digits never repeating, can in fact be generated by a very simple program. Since Wolfram believes that the secret of nature's complexity is the repeated application of simple

Of course, it is always possible that an advanced extraterrestrial intelligence decides to talk down to backward civilisations like ours. So, despite using efficient, pretty much random, signalling for its own communications, it chooses to broadcast just the kind of inefficient, patterned signal we would easily recognise. 'They might use a beacon that, by design, is energy inefficient, directional and easily detected by our current methods,' says Paul Shuch, Executive Director of the SETI League in New Jersey.* Max Tegmark of the Massachusetts Institute of Technology can imagine such an inefficient signal being broadcast for a deeply sinister reason. 'Say an extraterrestrial civilisation wanted to infect the computer network of a target planet with a devastating computer virus,' he says.

Wolfram thinks Tegmark and Shuch's scenarios are very unlikely, however. And for a simple reason. Any extraterrestrial intelligence that specifically tailors a signal for a civilisation like ours will be tailoring it for a civilisation that has developed radio communication but has not yet developed efficient radio communication. 'That's an awfully brief window in time,' says Wolfram.

What Wolfram means is that technological civilisations are not likely to be at such a stage for very long. So, at any particular time, there are likely to be very few civilisations in the Galaxy scanning the heavens for such a signal and, consequently, very little chance of the signal attracting the attention of its intended recipient.

Shuch, however, sees another possibility even if extraterrestrials do not

programs with complex outputs, it is conceivable that there is a natural process somewhere in the Universe that is currently employing the pi-generating program.
* The SETI League is an international band of radio and radio astronomy enthusiasts, dreaming of the day when one of them will catch ET phoning Earth. To this end, they are using satellite dishes and signal-processing software to listen out for extraterrestrial signals from their back gardens. (The SETI League, 433 Liberty Street, PO Box 555, Little Ferry, NJ 07643, USA; http://www.setileague.org.)

deliberately broadcast a signal of the type that SETI scientists are looking for. 'They might broadcast such a signal inadvertently,' he says. There may be emerging civilisations, pretty much like our own, he says, which waste energy by using inefficient, 'narrow-band', telecommunications. Inevitably, some of this electromagnetic chatter will leak away into space. 'Our leakage is currently detectable out to a distance of about 100 light years,' says Shuch. 'Theirs will be too.'

He concedes that most of our cosmic neighbours are likely to be far more advanced than us and so will not be using the kind of primitive communications we would recognise. 'Nevertheless, there are about 100 billion stars in our Milky Way,' he says. 'It stands to reason that there must be a few civilisations at our level, plus or minus a century or two.' Wolfram, however, sees the same problem as before. There are likely to be far too few civilisations out there at roughly our stage of development, making the chance of our detecting their electromagnetic leakage entirely negligible.*

However, deliberate broadcasts and inadvertent broadcasts may not be the only way that a technological civilisation gives away its presence. According to Seth Shostak of the SETI Institute in Mountain View, California, even a highly advanced society will want early warning of comets and asteroids that might be involved in a catastrophic collision with their home planet. 'And the best way to do this is to sweep interplanetary space with powerful radars,' says Shostak.

---

* Interestingly, a giant array of telescopes, being planned for operation in 2020, will be capable of picking up TV and radar broadcasts from planets round nearby stars. The Square Kilometre Array, which could be built in Argentina, Australia, South Africa or China, will consist of a cluster of dishes within five kilometres of each other, with outriders up to 3,000 kilometres away. With 100 times the collecting area of any existing telescope, just keeping the dishes working together as one will require data rates 100 times greater than the total UK Internet traffic in 2005!

Shuch points out that the radar signal necessary to locate city-sized chunks of rocks and ice flying through space is quite different from a telecommunications signal. 'By its very nature, it is inefficient and detectable from a long way off,' he says.

The Achilles' heel of this idea is once again that an extraterrestrial civilisation might use such a radar warning system for only a brief period of its history. Perhaps they will find a more effective means than radar to alert them to rogue comets and asteroids. Or perhaps they will vacate planetary surfaces for interstellar space. If the use of planetary defence radar is brief, it follows that the chance of SETI searches finding them is also slim.

Wolfram believes that, if extraterrestrials really want to communicate with us, there is a simple and effective way to get our attention. 'Why not simply use a big flash of light?' he says. 'It could be seen directly simply by looking up at the sky. There would be no need to rely on the recipients of the signal having constructed complicated radio telescopes.'

Extraterrestrial Artefacts

So much for picking up an ET signal. What about locating an extraterrestrial artefact? In 2005, Luc Arnold of the Observatoire de Haute-Provence in France suggested searching for extraterrestrial intelligence by looking for peculiar structures such as giant triangles orbiting nearby stars.[*] In the past few years, astronomers have detected planets which happen to pass in front of their parent star by the way

* See 'Transit Lightcurve Signatures of Artificial Objects' by Luc Arnold (*Astrophysical Journal*, vol. 627, p. 534). Also: http://xxx.lanl.gov/abs/astro-ph/0503580.

such 'transits' temporarily dim the star. Arnold points out that, if an unusual-shaped structure were orbiting a nearby star, it would dim the star's light in a way distinctly different from a planet. Such a structure might be part of the normal infrastructure of an alien civilisation – it might, for instance, be a giant collector of solar energy. Or it might be an artefact fabricated specifically to signal to civilisations around other stars in the Milky Way.

Wolfram, however, takes issue with Arnold. For pretty much the same reason he believes aliens would not use simple, patterned alien communication signals, he does not believe they would build simple, patterned artefacts. 'They are inefficient,' he says. 'They are very unlikely to be used by an advanced technology.'

To illustrate his point, Wolfram cites the example of trains arriving at a train station, say, every half an hour. According to Wolfram, even if we were to look down on the station from a great height – so high that we could see none of the details of the trains – we would still recognise their behaviour as artificial. The giveaway would be the regularity. 'However, this kind of regularity turns out not to be a necessary feature of a mass transit system,' says Wolfram. 'It is merely a reflection of the current primitive state of our transport technology.'

A more efficient transport system, according to Wolfram, would consist of small taxicabs shuttling back and forth on demand. Controlling them would admittedly require more computing power. But more computing power, according to Wolfram, is synonymous with greater efficiency. 'In the future, when our technological artefacts are continually communicating with each other and are able to carry out computations, the world will look a very different place,' he says. 'It will be far more complicated, far less patterned than it is today.'

At the present time, it is easy for us to distinguish between artificial objects and natural objects – for instance, between buses and trees.

Nature's artefacts are without fail more complex. 'But this will not always be true,' says Wolfram. 'In the future, all our technological artefacts will be as complicated as nature's – if not even more complicated.'*

It follows that the artefacts of an advanced civilisation will look as different from the giant space triangles envisioned by Arnold as it is possible to imagine. In fact, according to Wolfram, they will look pretty much like natural objects. This prompts an extraordinary question – to which Wolfram supplies an extraordinary answer. Could the stars be artificial? 'Yes, in the abstract,' says Wolfram. 'It is extremely difficult to rule out the possibility of them being built for a purpose.'

Extremely difficult but perhaps not impossible. Wolfram points out that, compared with biological entities which have evolved by natural selection, artefacts created by intelligence for some purpose tend to do fewer things which are not related to their principal purpose. A plane, for instance, does fewer spurious things than a bird. A bulldozer does fewer spurious things than the muscles of a human being. 'Artefacts created by intelligence are optimised for their purpose,' says Wolfram. 'On the other hand, it is inconceivable that a tree is optimised for its purpose.'

If Wolfram is right, in order to figure out whether or not an extra-terrestrial artefact or signal is of intelligent origin we must first guess its 'purpose', then see whether it does anything spurious to that purpose. But how do we guess the purpose of a signal or artefact? 'With great difficulty,' says Wolfram. 'In fact, I am not sure that it is possible at all.'

---

* The reason our artefacts will eventually be more complex than nature's is that natural selection is more restricted than human inventiveness. Rather than making radical changes to a creature, natural selection tends to make incremental changes – lengthening a neck or increasing the size of a brain. Human technological inventions are subject to no such constraint.

Deep Philosophical Waters

In fact, things are even worse than this. 'What do we even mean by "purpose"?' says Wolfram. 'This is a deep philosophical question.' And, according to Wolfram, the whole field of SETI is fraught with such philosophical difficulties. 'We are looking for "intelligence" but how do we even define "intelligence"?' he says. 'For that matter, how do we define "life"?'

The central qualities of life appear to be the ability to move about, compete for resources, reproduce, pass information from generation to generation, and so on. Living things also use characteristic 'biological molecules', including DNA, the genetic material; proteins, the scaffolding and chemical engines of cells; lipids, the molecules of cell walls; and carbohydrates, the fuel of life. 'But all of these things really are just characteristics of "life as we know it",' says Wolfram. 'They are specific properties of life on Earth – we have no idea of the general properties of life.'

So, what characteristic might fit the bill as a bona-fide general property of all life? Racking his brains, Wolfram can come up with only one: complexity. Unfortunately, if there is one thing Wolfram's exhaustive research over the years has revealed, it is that complexity of the degree seen in living things is not unique to living things. On the contrary, it appears to be exhibited in a vast range of other physical phenomena, from subatomic particle collisions to turbulence in a fluid to the circulation of the Earth's atmosphere. This discovery Wolfram has elevated to the status of a universal principle. Put crudely, the 'Principle of computational equivalence' says that systems of similar complexity are equivalent. In other words, a system which is as complex as a living thing – for instance, the circulation of the Earth's atmosphere – has exactly the same right to be classed as a living thing as you or me.

If defining life is difficult or impossible, defining intelligence is every bit as challenging. Take birdsong, says Wolfram. Zoologists have been able to say that a particular cluster of brain cells, or neurones, is responsible for a particular birdsong. They have then taken this as evidence that the birdsong, though undoubtedly complex, is not a manifestation of intelligence. 'The trouble with this argument is that each and everything we do – from reasoning to exercising creativity – is in principle traceable to discrete sets of neurones,' says Wolfram, 'What quality do we have that a bird does not have that makes us intelligent? The answer is not at all clear to me.'

Give up SETI?

With all the tremendous difficulties in recognising intelligent signals and artefacts – and even in recognising 'intelligence' at all – should the people doing SETI give up and go and work in petrol stations? Wolfram is not actually saying this. He is simply saying that SETI is going to be a lot harder than anyone imagines.

At present, SETI scientists look for patterned radio and optical signals not because this is the best thing they can do but because it is the only thing they can do. Two pioneers of SETI, Guiseppe Cocconi and Philip Morrison, said it all: 'The probability of success is difficult to estimate; but if we never search, the chance of success is zero.'* Wolfram understands this and that SETI scientists have little choice but to search in the way they do. Nevertheless, he thinks that, if anything worthwhile

---

* 'Searching for Interstellar Communications' by Guiseppe Cocconi and Philip Morrison (*Nature*, vol. 184, no. 4690, pp. 844–6, September 1959).

is to be achieved, the search strategy will have to be made a lot more effective.

At present, SETI scientists analyse the morass of data they collect from the sky with a technique known as 'Fourier analysis'. This picks out any patterns in the form of periodic signals. Wolfram, however, thinks people should also look for more subtle patterns in the data. He gives the example of testing whether or not a binary number – a sequence of 0s and 1s – is random. The first step is to see whether the number has as many 0s as 1s. A random number will have 50 per cent of each. The next step is to see whether there are any unusual blocks of 0s or 1s – say, nine 0s in a row. A random number will not have any such blocks. 'The point is that something like twenty tests are commonly used to check for randomness,' says Wolfram. 'In the same way, I think SETI signals should be analysed by a whole battery of tests. Relying on just one – the Fourier test – is really weak.'

Those in the SETI community have no problem with this. 'I agree wholeheartedly,' says Shuch. Shostak points out that there is already a more sophisticated approach now being used by SETI researchers in Bologna, Italy: 'It's called the Karhunen-Loeve Transform.' KLT promises to dig far deeper into the 'noise' than Fourier analysis. 'It stands a chance of detecting not only periodic patterns but also "aperiodic" patterns – ones which are not quite periodic,' says Schuch.

Nobody denies that SETI faces an enormous challenge. It is just that Wolfram thinks the challenge is even more enormous than anyone else does. But what if his pessimism is misplaced? What if, against all the odds, SETI scientists succeed in recognising an intelligent signal, an intelligent artefact, intelligence itself? Without doubt this will be a momentous day for the human race. However, Wolfram sees a problem even here. 'If we make contact with extraterrestrials, what could we possibly trade with them?' he says.

Search the Physical Universe or the Computational Universe?

The only thing Wolfram can think of is computer programs that can do useful or interesting things. He imagines the existence of an abstract 'computational universe' that contains all possible computer programs. When we find computer programs that do useful things – for instance, encrypt data or run spreadsheets – according to Wolfram, we are simply plucking these programs from the computational universe.

Even technological artefacts can be thought of as computer programs, according to Wolfram. In fact, he says that in the future there will be absolutely no distinction between computer programs that do things and physical artefacts that do things. This is because the electrons that shuttle about computers carrying out computations are the very same entities that glue together the atoms of matter. One day, Wolfram says, we will devise 'universal constructors' whose computational output will not simply be sequences of binary digits but actual physical artefacts – objects constructed from the Lego bricks of atoms. Arguably, nature already possesses such universal constructors.[*]

Wolfram points out that extraterrestrials will have access to the computational universe just like us. And this fact may severely limit what we can trade – what they can learn from us and what we can learn from them. 'After all, what could they possibly tell us except "We've done

[*] Cells take molecules such as amino acids as their input and, after carrying out the equivalent of a computation, produce as the output proteins, biomolecules that can perform a vast array of tasks from providing the scaffolding of cells to speeding up chemical reactions crucial to life. Nature's universal constructors, however, are constrained by natural selection and so cannot explore every nook and cranny of the computational universe. The universal constructors we may one day build will suffer no such constraint.

more computations than you and here are some good computer programs we've found"?' says Wolfram.

Of course, we may discover extraterrestrials who are far in advance of us and have explored much more of the computational universe. Leaving aside the issue of why they might want to give us the benefit of their experience when there is so little in it for them – how benevolent do we feel towards pigs, or bacteria? – might we not gain immeasurably from them? Yes, we might indeed. However, this is beside the point, according to Wolfram. The question is: How long is it going to take to find such an extraterrestrial civilisation compared with how long is it going to take to search the computational universe for their knowledge? According to Wolfram, there is no contest. Searching the computational universe is going to be quicker.

It has often been said that, if a monkey sat at a keyboard and hit the keys randomly, eventually it would write the complete works of Shakespeare. Of course, to stumble on the correct sequence of letters, spaces and punctuation, the monkey would have to sit at the keyboard for many times the present age of the Universe. Wolfram's key discovery, however, is that simple computer programs can create fantastically complex outputs. In other words, there exists a program which is enormously simpler than the complete works of Shakespeare that can generate the complete works of Shakespeare or, at the very least, generate an output in the style of Shakespeare.

The program for writing the complete works of Shakespeare will not be as difficult to find in the computational universe as the complete works of Shakespeare in the universe of all possible letter sequences typed by monkeys. Similarly, other interesting programs will be easier to find than might be expected. Since this is the case, why go to the trouble of searching the physical Universe for aliens to tell us about the computer programs they have found, asks Wolfram? Not only will we have to search 100 billion star systems in our Galaxy but it will be a darn sight harder than sitting at a computer on Earth and searching for the programs ourselves. 'It's more efficient for us to use a computer to simply search the computational universe,' he says.

Wolfram has already made a start at exploring the computational universe. In his monumental book, *A New Kind of Science*, he carries out a systematic search for all simple computer programs that have complicated and interesting outputs. According to Wolfram, the search for computer programs that do useful things in the computational universe is simply the latter-day equivalent of the search for useful commodities such as spices in the real world. Except, of course, it might take a lot longer.

Another logical consequence of Wolfram's view is that extraterrestrial intelligences themselves, just like human beings, will merely be particularly complicated computer programs. In other words, they will exist in the computational universe, as we will. So, instead of scanning the heavens for alien broadcasts, we could look closer to home. Much, much closer: your office PC or laptop. You could literally have an alien at your table! Of course, if we find ET in a computer, it will not be a flesh and blood alien – it will be a cyber version. (Unless of course we have acquired universal constructors and can build one from the digital blueprint.) But this does not mean that we cannot communicate with it. We could still converse with a virtual version of an alien civilisation, and learn plenty from the conversation.

Of course, all the knowledge we might gain from extraterrestrials – surely the principal reason for trying to contact them – is already ours for the taking. It is in the computational universe. We do not have to find extraterrestrials and ask them for it – unless, of course, we are feeling particularly lonely and would like someone to talk with. If we can find computer programs for aliens, we can find computer programs for what they know. We can cut out the middleman.

Wolfram is not saying do not search the physical universe for somebody to talk to. 'I'm merely saying we can look for extraterrestrial intelligence in the physical universe or we can look for extraterrestrial intelligence in the computational universe,' he says. 'But the task is a whole lot easier in the computational universe and there's a lot more there.'

10
The Billboard in the Sky

If the Universe was built and the builder wanted to leave a message, where would they have left it? In the 'afterglow' of the Big Bang

> Since the inflationary theory implies that the entire observed universe can evolve from a tiny speck, it is hard to stop oneself from asking whether a universe can in principle be created in the laboratory.
>
> Alan Guth, *The Inflationary Universe*, 1997

> Teacher told my parent that I am the slowest youngster in my class, but today I made a star in the third quadrant of kindergarten.
>
> James E. Gunn, 'Kindergarten', 1970

Today is the day. For two years the satellite has been drifting out in the darkness 1.5 million kilometres beyond the Earth. For two years it has been quietly measuring the cosmic background radiation, the cooled 'afterglow' of the Big Bang fireball which permeates every pore of space. Now the last gigabyte of data has been processed and the scientists who have devoted years of their lives to the mission crowd around a computer monitor at Mission Control.

In the expectant hush, fluorescent digits begin to scroll down the screen. To everyone's astonishment, there is a pattern. An unmistakable pattern. Someone has spotted numbers they recognise – the 'coupling constants' of the standard model of particle physics. One scientist whistles through her teeth, another gasps. This is no random signal from the beginning of time. It is a message from the Creator. And what it is saying is: 'This is how up I built the Universe.'

Surely such a scenario is pure science fiction? Not necessarily, according to two physicists in the US. Stephen Hsu of the University of Oregon in Eugene and Anthony Zee of the University of California at Santa Barbara asked themselves the question: if the Universe was built, and the builder wanted to leave a message for all the Universe's inhabitants, where would be the best place to leave it? 'In our opinion, there is only one place,' says Hsu.

When Hsu and Zee talk about the Universe being 'built', it needs to be stressed they are talking about 'someone' or 'something' creating the initial conditions that set the Universe in motion 13.7 billion years ago. They are not in any way subscribing to the 'intelligent design' claim that a Supreme Being individually designed the creatures that live on Earth. Furthermore, the someone or something that lit the touch paper of the Big Bang need not be a Supreme Being – God – but simply a superior being – a creature that is basically like us, only far more technologically advanced.

How to Make a Universe

One reason for thinking that the Universe could have been made by a superior being or superior beings is that, remarkably, we already know the recipe for making a Universe. It is called 'inflation'.

Inflation, a brief period of super-fast expansion, is widely believed to have occurred in the first split-second of the Universe's existence. As pointed out before, inflation was driven by the vacuum, which in the view of modern physics is not empty at all but seething with restless energy. According to proponents of inflation, the vacuum at the beginning of time was in a peculiar state quite unlike today's vacuum. In fact, it was so peculiar that it possessed repulsive gravity rather than the familiar attractive variety. This caused the vacuum in only a brief interval of time to balloon in volume by a staggeringly large factor. When inflation finally spluttered to a halt, the enormous energy contained in the, by now vastly inflated, vacuum had to go somewhere. And it went into making matter and heat – in short, it created the ferocious fireball of the Big Bang.

Remarkably, inflation could have been triggered by only a tiny 'seed' of matter – perhaps as little as a kilogram. This astonishing realisation has led to the much-repeated declaration by cosmologists that the Universe, with its countless galaxies and stars, is the 'ultimate free lunch'. Apart from the measly initial seed, all cosmic matter was born when the energy of the vacuum was dumped abruptly into mass-energy of subatomic particles.

The recipe for creating a universe is therefore clear: take a small seed of matter and subject it to the conditions that once triggered inflation in our Universe. Of course, there is a hitch – if there wasn't, someone would already have made a universe in the laboratory! The hitch is that, in order to create the peculiar state of the vacuum that leads to inflation, the seed must be squeezed to a tremendous density – equivalent to 10^{94} grams of matter crammed in a volume the size of a sugar cube.

Achieving such an ultra-compressed state of matter is clearly way, way beyond our technological capabilities. So, although we can see in principle how to make a universe in a laboratory, in practice we have no hope of doing so. Nevertheless, such an extraordinary feat may not be beyond the capabilities of a super-advanced civilisation. And this is the point. If a super-advanced civilisation could build a universe, then it stands to reason that such a civilisation could have built our Universe.

Why anyone would want to build a universe is a difficult question to answer. Guessing the motivations of super-advanced beings is a bit like a bacterium trying to guess the motivations of a human being. However, one possibility is that they would want to do it for exactly the same reason that we often do scientific experiments – simply to see what happens.

The cosmologist Edward Harrison has pointed out that, if the Universe was indeed created by superior beings, it could explain two puzzling features of our Universe. The first is why the laws of physics appear to be 'fine-tuned' for the existence of stars and planets and, ultimately, intelligent life like us. None of these things would be possible if, for instance, the forces of nature such as gravity were even a few per cent weaker or stronger than they actually are. Why, then, are the laws of physics just right for us to be here? Easy, says Harrison. If our Universe was made by superior beings, it would either be deliberately designed for life or else it may have inherited the conditions of its builders, who by definition, lived in a universe compatible with life.

A second very puzzling feature of our Universe was pointed out by Einstein and has never been satisfactorily explained. 'The most incomprehensible thing about the Universe,' he said, 'is that it is comprehensible.' Why, for instance, do we not live in a universe whose laws are so bafflingly opaque that we could never figure them out? Harrison has an answer. If our Universe was built by superior beings rather than an incomprehensible Superior Being, he says, it was created by comprehensible beings – beings far in advance of us but basically like ourselves. Intelligent but also intelligible. They made our Universe to be like theirs, and their universe was, in turn, understandable. How could it not be? They had to have enough understanding of it to manipulate it and make our Universe.*

* See my book, *The Universe Next Door* (Headline, 2002).

Where to Leave a Message

Hsu and Zee have no more evidence than Harrison that the Universe was made by a superior intelligence. However, this is not a concern to them. They are merely pointing out that, if the Universe was built, the builder may have wanted to leave a message to the Universe's inhabitants, saying perhaps: 'This is how I did it.' 'The question is: where would they leave such a message?' says Hsu.

Scientists and science-fiction writers have long speculated about where on our planet a message could be left in the safe knowledge that it would survive for a long time. Say, for instance, extraterrestrials came to the Solar System millions of years ago and wanted to leave a calling card for any intelligent creatures that might one day evolve on the promising third planet from the Sun. One possibility would be to inscribe a notice in an outcrop of solid rock somewhere on Earth. However, such a message would inevitably be worn away by the action of weather. A far better place to leave a message, argued science-fiction writer Charles Sheffield, is the genome of living things. Although natural selection causes DNA to continually change, or 'mutate', there are regions of DNA – junk DNA, or 'introns' – which stubbornly retain their identity from generation to generation. Such regions, Sheffield concluded, would be an ideal place for ET to leave a message for us. Just imagine. At this very moment the long-sought communication from the stars could be written in the heart of every fat red blood cell coursing through your veins!

DNA, however, is very specific to the Earth. It is not a viable message medium for a Creator who wished to advertise his handiwork to all inhabitants of the Universe. Nor is an artefact buried on the Moon like the mysterious 'monolith' excavated from Tycho crater in Arthur C. Clarke's *2001: A Space Odyssey*.

Other scientists, including the planetary scientist Carl Sagan, have suggested that a message from a super-advanced intelligence might be buried in the digits of a fundamental cosmic number such as pi, the ratio of the circumference to the diameter of a circle. To date many millions of digits have been computed, with not the slightest sign of any pattern in those digits. However, Sagan, in his science-fiction novel *Contact*, speculated that, after a billion digits, or a trillion, a pattern will eventually be found to emerge – and this will be a message from the Creator of our Universe.

Hsu and Zee accept that Sagan's idea is plausible. Unlike Sagan, however, they ask themselves where in the 'physical', rather than the abstract mathematical, Universe might a message be left – a message that could be seen from every star in every galaxy in the Universe? 'The answer is clear,' says Hsu. 'The cosmic background radiation.'

The Afterglow of the Big Bang

Ninety-nine per cent of all the Universe's photons are not in the light of stars and galaxies. They are instead tied up in the cosmic background radiation.[*] If we had eyes that were sensitive to short-wavelength radio waves rather than visible light, this would be glaringly obvious to us. We would see all of space glowing like the inside of a light bulb. 'And what we can see on Earth everyone else in the Universe can also see,' says Hsu.

The cosmic background radiation is the relic heat of the Big Bang fireball. It has been cooled so greatly by the expansion of the Universe

---

[*] See my book, *Afterglow of Creation* (University Science Books, Sausalito, California, 1996).

in the 13.7 billion years since the beginning of time that today it has an average temperature of only about -270 degrees Celsius. The reason it is still all around us today is simple – it was bottled up in the Universe and had absolutely nowhere else to go. Tune your TV between the stations and about 1 per cent of the 'static' on your screen will be the microwave relic of the Big Bang. Before it struck your TV aerial, the last time it interacted with matter was in the blistering inferno at the beginning of the Universe.

As noted earlier, microwaves arriving at Earth from very different directions come from regions in the Big Bang fireball that were not in 'causal contact' at the time the cosmic background radiation originated. This means the emitting regions were so far apart that light – the fastest thing in the Universe – could not have travelled between them. The importance of this for Hsu and Zee is that it means that nothing or nobody could have tampered with the cosmic background radiation in its entirety even at this early epoch – about 450,000 years after the Universe's birth. The only time a message could have been impressed on the fireball radiation was at the very beginning of the Universe – in its first split-second of existence.

In other words, only someone or something at the beginning – the builder of the Universe – could have left a message in the cosmic background radiation. And, after they had left it, no conceivable process could have erased it.

How to Encode a Message

But how exactly would the builder of the Universe encode a message in the cosmic background radiation? Hsu and Zee have a very specific idea. It requires a little diversion into the technicalities of the cosmic background radiation.

The average temperature of the cosmic background is 2.726 degrees Kelvin. But, as pointed out before, there are subtle variations in its temperature from place to place in the sky – 'hot spots', which are ever-so-slightly warmer than average, and 'cold spots', which are ever-so-slightly cooler than average. These arise because the matter in the Big Bang was ever so slightly lumpy. The lumps have since been magnified by the remorseless action of gravity to make the galaxies including our own Milky Way.

The hot spots and cold spots in the cosmic background radiation occur at all sizes. For instance, there are big blotches which stretch across much of the sky and, superimposed on these, smaller goose pimples. To make sense of everything, astronomers like to separate out the different components, breaking up their 'temperature map' of the microwave sky into what they call 'multipoles'.

The simplest multipole is the 'dipole'. This is simply one huge hot spot and one huge cold spot. Actually, this has nothing whatsoever to do with the Big Bang. Rather, it is the temperature variation caused by the motion of the Milky Way which is flying at hundreds of kilometres a second through the photons of the cosmic background radiation. This makes the afterglow of the Big Bang appear hotter in the direction the Milky Way is flying and colder in the opposite direction.

After the dipole temperature variation, the second simplest multipole is the 'quadrupole'. The best way to think of this is as two dipoles – that is, two hot regions and two cold regions. Next comes the third simplest multipole, the 'octupole', which consists of three dipoles – that is, three hot regions and three cold regions. You get the idea. The simplest multipole components correspond to the biggest blotches and the more complicated multipoles to the smallest freckles.

Now, associated with each multipole is a maximum temperature variation – the difference between the coldest and hottest region. Astronomers call it the 'amplitude'. For instance, the amplitude of the dipole variation in temperature is several hundred times bigger than that of the quadrupole. 'It is these amplitudes that we believe are the ideal places for the Creator of the Universe to lodge a message to the Universe's occupants,' says Hsu.

Hsu and Zee point out that, within the foreseeable future, scientists will measure not only the amplitude of the dipole, quadrupole and octupole variations in the cosmic background radiation but the amplitude of the first 10,000 multipoles. This will provide them with 10,000 unique numbers describing the cosmic background radiation. Exactly how much information can be extracted from these 10,000 numbers depends on how accurately it is possible to measure them. Hsu and Zee estimate that ten bits of information could be encoded in each amplitude, making a grand total of 100,000 bits. Not a lot compared to the gigabits that can be stored on a PC's hard drive but enough to leave a potentially priceless message for the Universe's occupants.

How exactly would the builder of the Universe fix those amplitudes? Well, the lumpiness of matter is thought to have itself originated in the first split-second of the Universe when the vacuum on the microscopic scale was seething like the surface of water boiling in a saucepan. Hsu and Zee speculate that anyone who was able to manipulate these undulations of space could imprint indelible marks on the cosmic background radiation.[*]

---

[*] Specifically, Hsu and Zee propose a mechanism in which the self-interactions of the 'inflaton' – the field responsible for inflation – caused the variations in lumpiness. This proposal relies on physics – quantum field theory – that physicists consider well understood. It requires only that the superior being fine-tuned the inflation dynamics.

So Much for the Medium, What About the Message?

What might the Creator put in the 100,000 bits available to it? That is of course a difficult question. However, Hsu and Zee speculate that they might tell us how the Universe has been built. As far as we are aware, four 'fundamental' forces orchestrate everything that happens in the cosmos. As discussed before, these forces arise from even more fundamental entities called gauge fields. The gauge fields, in turn, can be described by 'matrices' – nothing more than tables of numbers.* It is these numbers that Hsu and Zee think the Creator might encode in the cosmic background to tell us how creation is put together. 'In effect, they would be saying "Hey guys, the Universe is governed by gauge fields, with the following structure . . ."' says Hsu.[†]

Some physicists believe, however, that the most fundamental theory describing reality is 'string theory', which views the fundamental entities – the building blocks of all matter – as ultra-tiny 'strings', vibrating in ten-dimensional space. If string theory is the correct description of our Universe, then the Creator might have left the details of string theory in the cosmic background. 'Of course, these are just specific suggestions,' says Hsu. 'Perhaps our collective scientific mind is still too puny to guess what the message on the billboard in the sky might read.'

Maybe the Creator simply signed their creation with the equivalent of 'I woz 'ere'!

Hsu and Zee urge that, when more accurate cosmic background data becomes available, it should be analysed carefully for possible patterns impressed on it by the builder of the Universe. 'We believe we have

---

* See Chapter 7, 'Patterns in the Void'.
† Of course, Stephen Wolfram believes that it may be impossible for us to recognise any message from a superior intelligence! See Chapter 9, 'An Alien at My Table'.
‡ See Chapter 3, 'Yoga Universe'.

raised an intriguing possibility,' says Hsu. 'Searching for a signal in the cosmic background radiation may be even more fun than the search for extraterrestrial intelligence.'

11
The Never-Ending Days of Being Dead

Can life survive for ever in the Universe? Yes! – as long as it steers the Universe along a very special path.

They will have time enough, in those endless aeons, to attempt all things, and to gather all knowledge ... no gods imagined by our minds have ever possessed the powers they will command ... But for all that, they may envy us, basking in the bright afterglow of creation; for we knew the Universe when it was young.
Arthur C. Clarke, *Profiles of the Future*, 1962

'We are alive,' said Lucinda and at that moment she felt herself to be what she said. 'We are alive and on the very brink of eternity.'
Peter Carey, *Oscar and Lucinda*, 1988

You've had a long life but, finally, your time has come. If you were a wit like Oscar Wilde, you would say something amusing like: 'Either these curtains go or I do.' But you are suffused by such an awful tiredness that you can barely think, let alone speak. You no longer have the strength to fight on. This is it. As your eyelids begin to fall, coming down like metal shutters on your life, the hubbub of the world fades to a distant murmur. You draw one last breath ...

... and it is summer and you are young again. Your favourite dog – the one you loved so much as a child and thought you would never see again – has knocked you to the ground and is licking your face furiously.

Through tears of joy, you see your father and mother – long dead – standing over you. They are young – just as they were when you were ten years old – and they are laughing and stretching out their hands to you.

What is happening? Have you died and gone to Heaven? Not exactly. You've been resurrected as a simulation on a computer at the end of time!

Surely, such a scenario is pure fantasy? You would be forgiven for thinking so. But, according to a prominent American physicist, an extraordinary fate like this awaits each and every one of us after we die. His name is Frank Tipler and he has come to this outrageous conclusion not for any theological reasons but after applying the formidable armoury of modern physics to a single, profound question – Can life survive for ever in the Universe?

Life Needs Energy

The first scientist to seriously address this question was the Anglo-American physicist Freeman Dyson in 1979.[*] His immediate problem was to define 'life', a thorny and controversial issue which is to this day hotly debated. Dyson settled on the idea that a living thing is a 'processor of information'.

The advantage of this definition is that it is general. For all we know, life in the far future may be implemented not in the wetware of biology but in the software of computers – or, more likely, in some form beyond anything we can currently imagine. But regardless of how life is implemented, argued Dyson, it will still have to process information. The question of whether life can go on for ever in the Universe can

[*] *Reviews of Modern Physics* (vol. 51, p. 447, 1979).

therefore be turned into another, more tractable, question. Can information processing go on for ever? The answer, realised Dyson, depends crucially on the long-term fate of the Universe.

'Prediction is always difficult, especially of the future,' warned the Danish physicist Neils Bohr. However, in the case of cosmology – the science of the Universe – there are grounds for optimism. Currently, the Universe is expanding, its constituent 'galaxies' flying apart from each other like pieces of cosmic shrapnel in the aftermath of the Big Bang. In the ultra-long-term future there are two possibilities.

One possibility is that the expansion of the Universe will eventually run out of steam and the Universe will then embark on a phase of runaway contraction. If we live in such a 'closed' universe, we can expect it to shrink all the way back down to a Big Crunch, a sort of mirror image of the Big Bang in which all the galaxies are piled up on top of each other in a great cosmic smash-up. The other possibility is that the Universe will continue to expand for ever, its constituent galaxies perpetually receding from each other into the empty cosmic night. In such an 'open' universe, the fate of everything, to coin a phrase from T. S. Eliot, will be to end 'not with bang but with a whimper'.

The thing that determines whether the Universe is open or closed is the amount of matter it contains. If there is more than a critical threshold of matter, its combined gravity will be enough to brake and eventually reverse the expansion. If there is less than the threshold, nothing will ever stop the expansion. In 1979, when Dyson considered the question of whether life could survive for ever in the Universe, it looked very much to astronomers as if there was insufficient matter tied up in stars and galaxies and clouds of gas drifting out in space to close the Universe. It was therefore in the context of an open universe that he posed his question: Can information processing go on for ever?

The processing of information involves what physicists refer to as 'work'. In the case of an everyday PC, for instance, work must be done to flip a transistor on a microchip from a state representing the binary digit '0' to a state representing a '1', and vice versa. Work is possible, however, only if there exists an energy difference. For instance, a difference in electrical energy – characterised by a voltage difference – drives the electrical current that flips the state of a transistor. Similarly, a difference in heat energy between the Earth's equator and the poles – characterised by a temperature difference – drives the planet's weather systems.

In the Universe as a whole, the ultimate driving force behind all activity is the temperature difference between different regions – specifically, between the stars like the Sun, which are tremendously hot, and interstellar space, which is tremendously cold.

If the Universe is eternally expanding, however, the stars will eventually burn out and even the feeble heat that criss-crosses space will be diluted by being smeared throughout greater and greater volumes of emptiness.* The overall effect will be to iron out any differences in temperature that exist between different parts of the Universe so that, eventually, no further work will be possible and all cosmic activity will die away. Ever since the nineteenth century, when this miserable cosmic end-state was first recognised as a distinct possibility, physicists have referred to it as the 'heat death' of the Universe.

On the face of it, an ever-expanding universe, heading for inevitable heat death, would appear a very unfavourable arena for the processing of information – and therefore life – to continue for ever. Appearances, however, can be deceptive. Surprisingly, Dyson's calculations showed that, even in such an unpromising universe, life can survive indefinitely. Mind you, its survival is at a considerable cost. Life must continually slow itself

---

* The feeble heat that fills space is mostly the left-over heat of the Big Bang, cooled to a chilly -270 degrees Celsius by the expansion of the Universe in the past 13.7 billion years.

down, even entering a state of hibernation for extended periods of time.

If life gets ever slower, constantly scaling down its energy needs, even the meagre temperature differences between different regions of the Universe can be enough to keep it ticking over. For a creature condemned to live in the far future of such a universe a single thought might take a million years, a billion years, or even more. It would not matter, however, how long it took. After all, if there is one resource that is available in inexhaustible abundance in an eternally expanding universe it is time.

But, as Dyson discovered, slowing down may not be enough on its own to ensure the perpetual survival of an organism. At times there may be insufficient energy for even a single thought. This is where hibernation comes in. By entering a period of suspended animation, a being can wait patiently until enough of a temperature difference has built up in the external universe to drive its next thought.

By a clever combination of slowing down interspersed by periods of hibernation, life can survive into the infinitely far future. The active word here is of course 'survive'. By no stretch of the imagination can life be said to flourish. It merely clings desperately to existence. 'And who would want to live that way?' says Tipler.

Another difficulty with Dyson's survival strategy is connected with the 'red shift'. As the Universe expands, light – the only means by which an entity can transmit its thoughts from one part of itself to another – becomes stretched out, or red-shifted.* When light is stretched in this

---

* The red-shift effect can be pictured by drawing a wave on a balloon. If the balloon is inflated, the wave is stretched. This is exactly what happens to a light wave as the fabric of space expands. Physicists characterise light by its 'wavelength', which is a measure of the distance between successive crests of the wave. And, since red light has the longest wavelength of any visible light, stretching the wavelength of light shifts it towards the red end of the 'spectrum', *red-shifting* it.

way, it loses energy. To compensate for this an organism therefore needs even more energy for its internal signalling. In a universe already severely strapped for energy, the red shift makes things worse, pushing up still further the energy requirements for the processing of information.

The red-shift problem – not to mention the general unattractiveness of Dyson's scenario – set Tipler wondering whether there was any other kind of universe in which life could survive for ever. Since Dyson had considered an open universe, Tipler considered the only other possibility: a closed universe.

Can Life Survive For Ever in a Closed Universe?

At first sight, it seems patently obvious that life cannot go on for ever in a closed, shrinking, universe. A closed universe, after all, eventually comes to an end. However, it is important to make a distinction between the external time in such a universe – which undeniably is in limited supply – and the internal, or 'subjective', time which is experienced by a creature living in the universe. The two may be quite different.

That subjective time is an elastic concept is common human experience. For an eighty-year-old, a year appears to fly by, while for an eight-year-old a summer's day seems to stretch for ever. A child lives its life perpetually on the brink of eternity, blissfully unaware that it will one day die. However, it is not necessary to be unaware of death to experience time in an altered manner. A similar effect might be achieved if it were possible to change the speed at which thoughts are processed. In Dyson's heat-death universe, for instance, a creature ensures its perpetual survival by drastically slowing down its thought processes. In the time it takes it to yawn and slip back into hibernation, trillions of years may have marched on by in the external universe. Of course, this slowing down of information processing is the complete opposite of what is necessary to extend subjective time in a closed universe. Here, external time is severely limited by the looming Big Crunch and a creature must speed up rather than slow down its thought processes in order to squeeze as much as possible out of every second remaining before the cosmic curtain call.

Whether or not this is possible depends crucially on the temperature difference between different parts of the Universe since this is the driving force behind all cosmic activity. In the case of a heat-death universe, the temperature difference dwindles to next to nothing, causing cosmic activity, including information processing, to all but cease. If, however, information processing is to speed up, the temperature differences throughout the Universe must grow bigger.

At first sight, a closed Universe heading for the inevitable Big Crunch seems promising. After all, things get hot when they are compressed into a small volume, as anyone who has squeezed the air in a bicycle pump knows. So, as the universe shrinks, more and more heat energy will become available. In fact, if the universe shrinks all the way to zero size, an infinite amount of energy should become available.* It would seem that in a closed Universe there is an embarrassment of energy for information processing.

The problem is that it is not energy that is required to do work, including process information – it is an *energy difference*. And, in a shrinking universe, everywhere heats up at the same rate. So, no matter how hot it gets, it gets hot everywhere pretty much equally and no

---

* Actually, physicists expect quantum theory to prevent a Big Crunch universe from shrinking all the way to a single point, or 'singularity'. Unfortunately, because they do not yet possess a quantum theory of gravity, they do not quite know how this fate is to be avoided.

appreciable temperature difference develops. Consequently, there is no scope for the speeding up of the processing of information – and no scope for a creature to live an enormously long subjective time in the limited time before the Big Crunch.

It would appear that, despite its abundance of energy, a closed universe is an even worse place for life to cling to existence than an open, heat-death universe. Perhaps this is obvious. A closed universe, with the Big Crunch presiding over the end of time like a black widow spider, never seemed very promising. But it is possible to be too hasty. What exactly is it about a closed Universe that is the real problem?

The real problem, according to Tipler, is that such a universe shrinks at the same rate in all directions. This is the reason why no appreciable temperature differences ever arise. And this is the reason why information processing cannot speed up. But does it have to be this way? What would happen if the Universe did not shrink at the same rate in all directions? This is the remarkable possibility considered by Tipler.

The Omega Point Universe

In Tipler's universe, space shrinks faster along two directions than along the third. Imagine a terrestrial globe that tightens its belt – shrinking in on itself at the equator far faster than it shrinks between the poles. In a universe behaving in this way, the stuff that shrinks fastest – the material in the plane of the equator – gets hotter than the stuff that does not – the material out by the poles. And this difference in temperature just keeps on growing.

The bigger and bigger temperature difference drives ever more cosmic activity, enabling information to be processed faster and faster.

Eventually, if the Universe dwindles down to a dimensionless point, the temperature difference will become infinite, which means it will be possible to process information infinitely fast!

Surely, this is too good to be true? Certainly, there are possible problems with this scenario (not least how such a peculiar universe could come about!). For one thing, as the temperature goes up, more energy is needed to store each single 'bit' of information. After all, by definition, information can exist only if it is noticeable. And that requires it to have a temperature above the Universe's average temperature, which is of course sky-rocketing.

Another problem with Tipler's universe is that the processing of information inevitably generates waste heat, which must somehow be got rid of. In the case of a PC, the waste heat is commonly expelled into the air by a cooling fan. This warms up the surroundings, reducing the temperature difference between the computer and its environment. Such an evening-out of the temperature is small for a PC. However, for a searing-hot universe in which enormous quantities of heat must be dissipated, it might seriously slow down the processing of information.

At first sight, the problems of getting rid of waste heat and finding enough energy to store the information being processed appear serious. However, as the Universe shrinks, the temperature difference which drives information processing continually goes up. The key question is therefore: does it go up fast enough to compensate for these problems? Tipler has done the calculations, and the remarkable answer is, yes!

Incredibly, the laws of physics permit a future universe in which the energy for information processing goes up without limit. In such a universe, the amount of energy available grows at a faster rate than the time left in the universe shrinks. So, although any creature that adapts to think faster and faster will have less and less time left to do their thinking, this will be entirely compensated for by the fact that they can do more and more in that time. From their subjective point of view, the imminent end of the universe will appear to recede until it is an infinite distance away in their future. There will literally be time enough for an eternity of living before the curtain comes down at the end of the universe. 'Time yet for a hundred indecisions. And for a hundred visions and revisions. Before the taking of toast and tea', in the words of T. S. Eliot.[*]

If our world was to end in a day but someone were to find a way of speeding up our lives so we could fit a whole lifetime of experiences in a single hour, in a single minute, what would the end of the world matter to us?[†] Similarly, of what consequence would the end of the Universe be to creatures who can squeeze an infinite number of lifetimes in the remaining seconds, remaining nanoseconds – creatures who, in the ever-shrinking time that is left, can live for a subjective eternity?

The fact that the laws of physics permit a future universe in which life can potentially survive for ever is a miraculous thing to behold. In fact, Tipler believes it is so remarkable that it can be no accident. It must mean that life has a crucial, though mysterious, role to play in the Universe. Tipler has coined a name for the extraordinary processing frenzy in which such an eternal universe ends. He calls it the 'Omega Point'.[‡]

---

[*] 'The Love Song of J. Alfred Prufrock'.
[†] Arthur C. Clarke explores a scenario like this in his story 'All the Time in the World' (*The Other Side of the Sky*, VGSF, 1987). A petty criminal is approached by a mysterious man who gives him a bracelet which speeds up his personal time so that the outside world appears to creep by in ultra-slow motion. All the criminal has to do – for a ridiculous payment of millions of dollars – is to use the bracelet to steal a list of major world treasures. Only too late does he learn that the mysterious man is an alien who knows that the Sun is about to go nova and has come to loot the Earth before the planet is incinerated.
[‡] The term is actually borrowed from the French theologian Pierre Teilhard de Chardin.

The Omega Point universe could never come about naturally. The only way to create a universe that shrinks at different speeds in different directions is artificially. It must be 'engineered'. Life must grab the Universe by the scruff of the neck and steer it in the desired direction. It must exert a controlling influence over the long-term fate of the Universe.

To say this is a staggering challenge is a bit of an understatement. Before even asking how it might be done, there is another, more basic question: Why do it?

Why Engineer an Omega Point Universe?

The answer, according to Tipler, is self-evident: 'Because life can do nothing else.' Ever since the emergence in the primeval slime of the first self-replicating molecule – the ancestor of all terrestrial organisms – life on Earth has been driven by one overriding impulse: the need to perpetuate itself. And the only way life can perpetuate itself indefinitely in our Universe is if it finds a way of driving things along the path that leads to the Omega Point.

As Dyson has observed: 'It is impossible to calculate in detail the long-range future of the Universe without including the effects of life and intelligence. It is impossible to calculate the capabilities of life and intelligence without touching, at least peripherally, on philosophical questions. If we are to examine how intelligent life may be able to guide the physical development of the Universe for its own purposes, we cannot altogether avoid considering what the values and purposes of intelligent life may be. But, as soon as we mention the words value and purpose, we run into one of the most firmly entrenched taboos of twentieth-century science.'

Tipler, like Dyson, is unafraid to confront the taboos. He believes that life, driven by its unwavering instinct for survival, will spread to fill the entire Universe, at which point it will inevitably contemplate the ultimate engineering project – the steering of the Universe down to the promised land of the Omega Point.

If this sounds like madness, think again. According to Tipler, the first step on the long road to the Omega Point is pre-ordained. It is an inescapable consequence of the survival impulse. 'Provided that humans do not wipe themselves out in some global catastrophe', he says, 'our descendants will one day leave for ever the cradle of the Earth and spread among the stars.'

This abandonment of Earth will not be a matter of choice, says Tipler. It will be forced on our descendants. In about five billion years, the Sun will have exhausted the hydrogen fuel deep in its core. By then, it will have puffed up into a monstrous, super-luminous 'red giant', pumping out more than 10,000 times the heat it does today. If this bloated star does not completely swallow our planet – and it will definitely envelop the close-in worlds of Mercury and Venus – it will certainly reduce the Earth to a burnt and blackened lump of slag.[*][†]

---

[*] The Earth will actually become uninhabitable in much less than five billion years. This is because the Sun is slowly getting hotter as it burns through its hydrogen fuel (since its birth 4.56 billion years ago, it has brightened by about a third). As the Sun continues to get hotter, it will drive out the 'greenhouse gas' carbon dioxide from chalk cliffs, which will lead to yet more global warming, which will in turn cause the oceans to gradually boil away. The Earth's water vapour will be destroyed by solar ultraviolet light when it rises to the top of the atmosphere. This, according to Juliana Sackmann of the California Institute of Technology in Pasadena and Arnold Boothroyd of the University of Toronto, will, within a mere billion years, leave the planet an uninhabitable desert.

[†] Many astronomy books say the Earth will be swallowed by the Sun which, as a red giant, will balloon out almost to the orbit of Mars. However, Juliana Sackmann's

The death of the Sun sets a critical deadline for our far-future descendants. Long before a blood-red orb swells to fill half the terrestrial sky, they must have vacated their home planet for the stars.

The crossing of interstellar space presents an immense and daunting challenge. The environment between the stars is unimaginably harsh. Creatures made of flesh and blood can survive it only if they are shielded from the hard vacuum, from the shattering cold and from the lethal sleet of cosmic ray particles which permeate space and can blast apart the fragile strands of DNA.

Tipler thinks the necessary degree of shielding is unattainable. He is therefore pessimistic that humans can tough it out in the abyss between the stars. 'Flesh-and-blood is simply too fragile for the rigours of interstellar space,' he maintains. 'And, by the time our descendants are ready to leave the Solar System, they will have long realised this.'

According to Tipler, there is only one way our descendants can possibly cross interstellar space – in the guise of machines. 'First they will have to "download" their minds into computers,' he says. 'Minds implemented in machinery will be much easier to harden against the

team has pointed out that, although the Sun will certainly get to the Earth's orbit, when it does the Earth will not be there! It is all down to the fact that red giants lose material at a terrific rate via their 'stellar winds'. A less massive Sun will also have weaker gravity with which to hold onto the Earth, so the Earth will gradually move away. By the time the Sun reaches the Earth's current orbit, it will have only 60 per cent of its present mass and the Earth will be 70 per cent farther away, so the planet will probably escape being gobbled.

A team led by Mario Livio of the Space Telescope Science Institute in Baltimore, however, points out there is a competing effect. The Earth raises a 'tidal bulge' in the Sun, which it will try to drag around with it as it orbits. As a consequence, the Earth will 'spin-up' the envelope of the Sun while it slows and moves inward. The rate at which the Earth is sapped of orbital energy depends crucially on how viscous is the stuff of the Sun's envelope, which nobody knows well. Currently, therefore, it is not possible to tell which of the two effects will win and whether or not the Earth will be gobbled.

conditions of interstellar space than minds implemented in jelly and water.'

If Tipler is right, the ships that will one day stream outwards from the dying Earth, like so many dandelion seeds scattered to the wind, will contain steel and plastic not flesh and blood.

The aim of this diaspora will be to seek out and colonise inhabitable worlds around other suns. However, the new worlds – other Earths – can never provide much more than a temporary respite from the exigencies of survival. Other suns will grow old and die, just like our own. Faced with this inescapable fact of cosmic life, our far-future descendants will have no choice but to keep on moving, for ever discarding used-up planets before they can be scorched and blackened by their dying suns. In this way they will spread inexorably from one end of the Galaxy to the other.

Our Milky Way is a great, ponderously turning, pinwheel of stars, crammed with a few hundred billion suns and spanning about 150,000 light years. This is almost 40,000 times the separation between Sun and its nearest stellar neighbour, Alpha Centauri, so the challenge of spreading throughout the Galaxy is of an altogether more formidable character than merely reaching nearby stars. Tipler, typically, is unfazed. He believes there is a sure-fire way of colonising the Milky Way – and relatively quickly. It involves exploiting a device first envisaged by the Hungarian-American mathematician John von Neumann.

A 'self-reproducing von Neumann probe' is a cross between a starship and a robotic factory. As the first step in the colonisation of the Galaxy, a large number of such probes would be despatched to nearby stars. On arrival at their target planetary systems, they would land on planets or moons or asteroids and set about using the available resources to build copies of themselves. Once completed, the copies would depart for nearby stars, where the whole process would repeat itself. In this way, von Neumann probes would multiply and spread throughout the Milky

Way, infecting billions upon billions of planetary systems like some unstoppable galactic virus.

How long it will take these robot descendants of the human race to reach every last nook and cranny of the Milky Way will clearly depend on how fast they travel. According to Tipler, a speed close to the cosmic speed limit itself – the speed of light – is possible. Boosting a probe close to such a tremendous speed will of course not be easy. However, Tipler believes it is possible by exploiting the ultimate fuel – antimatter.[*]

Antimatter's key characteristic is that, when it meets matter, 100 per cent of its mass-energy is instantly converted into other forms of energy such a heat. An H-bomb, by contrast, turns less than 1 per cent of its mass-energy into heat energy. It is antimatter's ability to pack the biggest punch possible for a given weight that makes it the fuel of choice for interstellar travel. The only problem is obtaining enough of the stuff. Despite the best efforts of physicists, they have so far managed to accumulate no more than a billionth of a gram of antimatter. Tipler, however, thinks this is merely a technological problem, and that almost certainly it will be solved in the fullness of time.

Tipler envisages antimatter-powered von Neumann probes travelling at about 90 per cent of the speed of light. These will take about twenty million years to colonise the Galaxy. Though admittedly a very long time in human terms, this is a mere blink of the eye in the life of the Galaxy – less than a tenth the time it takes the great flywheel of stars to turn once on its axis.[†]

[*] Every subatomic particle has an associated antiparticle with opposite properties such as electrical charge. For instance, the negatively charged electron is twinned with a positively charged antiparticle known as the positron. When a particle and its antiparticle meet, they self-destruct, or 'annihilate', in a flash of high-energy light, or gamma rays. Antimatter is the term for a large accumulation of antiparticles.

[†] Actually, the pinwheel of our Galaxy does not rotate like a solid body. The speed at which stars at a particular distance from the centre of the Galaxy orbit the centre

With the Milky Way colonised, the next obvious target for the von Neumann probes will be the 'Local Group', the sparse cluster of galaxies dominated by the great spirals of the Milky Way and the Andromeda Galaxy. According to Tipler, this could be overrun in a mere billion years. And, with the Local Group conquered, our robot descendants will stand at the very brink of known space, contemplating the great black gulf beyond and the galaxy clusters marching away into misty invisibility. Their next challenge – pre-ordained by the survival imperative – will be nothing less than the colonisation of the entire observable Universe.

It goes without saying that visiting every star in every galaxy in the entire observable Universe is a long-term project! Tipler estimates that it will take about twenty billion years, which is slightly longer than the Universe has currently been in existence.

Of course, the tacit assumption in all this discussion is that our space-faring descendants have things totally their way and encounter no serious opposition among the teeming stars and galaxies. But what if other intelligent races are abroad in the Universe? Surely, these too will spread among the stars, impelled by exactly the same survival imperative as the human race? Colonising the Universe might not be an option – it may already be occupied.

Tipler, however, is confident that this is not the case. He believes that we are the first intelligent to race to appear on the scene, certainly in our Galaxy, and he bases this claim on a straightforward observation. The Earth does not appear to have been visited by extraterrestrials.

Tipler's logic is simple. Any civilisation that gains a space-faring capability can clearly build their own von Neumann probes. In twenty

depends on how much material there is inside their orbit pulling on them with its gravity. For this reason, the orbital speed of stars is different at different distances. At the Sun's distance from the centre – about 27,000 light years – the stars go round on the galactic merry-go-round about once every 230 million years.

million years or so, they can spread to all the stars in the Galaxy – including our Sun. But they are emphatically not here! 'Where is everybody?' as the Italian–American physicist, Enrico Fermi asked.[*]

The only way to resolve the 'Fermi paradox', according to Tipler, is to accept that intelligence has arisen on Earth before it has arisen elsewhere. Incredible as it seems, we are the first. 'Sometimes I think we are alone, sometimes I think we are not,' said Buckminster Fuller. 'Either way, the thought is staggering.'

Not only is it a staggering thought that we might be totally alone in the vast Universe, it is a deeply sad one too. The human race, like the last person alive after a global catastrophe, is destined never to find anyone else to talk to, never to find anyone else with whom to share a single one of its experiences. But cosmic loneliness is not the only curse of being the first intelligence to emerge. A huge burden of responsibility comes with it. 'It will be up to us, and us alone, to ensure the survival of life into the eternal future,' says Tipler.

By the time our descendants have spread to fill all the stars and galaxies in the Universe, the Universe will be more than twice as old as it is today. With so many more stars burnt out, it will be a darker, colder place. The dying of the light will serve only to underline the awful dwindling of options open to our far-future descendants. They will die out, along with the stars all around them – unless they can rise to the ultimate challenge. That challenge will be to gain control of the Universe and force it, kicking and screaming, towards the Omega Point,

---

[*] Fermi, the creator of the first nuclear reactor in 1942, asked the question at the Los Alamos Laboratory in New Mexico in the summer of 1950. At the time, he was having lunch with a handful of physicist friends including Edward Teller, the 'father of the H-bomb'. Knowing Fermi's genius, Teller and the rest immediately realised he had asked a profound and troubling question. See *If the Universe is Teeming with Aliens – Where is Everybody? Fifty Solutions to the Fermi Paradox and the Problem of Extra-terrestrial Life* by Stephen Webb (Copernicus, New York, 2002).

with its seductive promise of survival for subjective eternity.

But how can such an engineering project – the most ambitious conceivable – ever be carried out? Remarkably, there is a way.

Chaos in the Universe

Almost certainly, it will be necessary to shunt large amounts of matter around the Universe, boosting the density in some places and lowering it in others. Think of great clusters of galaxies being moved from place to place like chess pieces on a gargantuan chess board. This will be astroengineering on an unimaginably heroic scale. But it will be needed to ensure that the Universe collapses in on itself in the necessary way, shrinking faster in one direction than in the others.

For a while, as the Universe shrinks, everything in the cosmological garden will be rosy. The ever-growing temperature differences will drive ever-more information processing. Nothing will appear to stand in the way of the attainment of the Omega Point. All, however, is not quite as straightforward as it seems.

Matter today is not smeared completely smoothly throughout the Universe and nor will it be in the far future. Inevitably, there will be slightly more material in one place than another. The trouble with this is that, as the Universe shrinks ever close to zero volume, even tiny irregularities in the matter distribution will become grossly magnified. And this is not all. According to Einstein, matter warps space. So the gross irregularities in the matter distribution will cause gross distortions in the fabric of space-time as well. Those distortions will change from instant to instant. Like a piece of toffee, the Universe will be stretched one way one moment and another the next.

But this is not the worst of it. As John Barrow of Cambridge University has shown, the distortions in space-time in the far future of the Universe are 'chaotic'.

It is characteristic of chaotic systems that their long-term behaviour is unpredictable. A good example is the Earth's weather system. It is impossible to forecast with any reliability what it will be like at a given location on the planet more than a week or so in advance. In the same way, as the Omega Point is approached, it will be impossible to predict the chaotic fluctuations in the distortion of space-time. With every instant they will become more unstable, more violent.

All is not lost, however. Although the long-term evolution of the Universe is chaotic and unpredictable, paradoxically it presents life with an opportunity.

The source of the unpredictability of the weather is the fact that chaotic systems are fantastically sensitive to initial conditions. Even a tiny difference in the state of the atmosphere on one day will lead to a dramatically different state a month later. As chaos researchers often point out, the beat of a butterfly's wings in one part of the world can eventually spawn a hurricane in another part of the world, a feature known as the 'butterfly effect'.

If something as insignificant and trifling as the fluttering of a butterfly's wings can in the fullness of time spawn something as tremendous and awe-inspiring as a hurricane, it opens up a remarkable possibility. It might be possible to change the weather in a major way merely by modifying the atmosphere in a minor manner – for instance, by tweaking the temperature of a small patch of sea, covering perhaps a few hundred square kilometres. With the aid of such 'weather control', we might one day be able to deflect a hurricane from its deadly path and spare a major city.

As with the weather, so with the Universe. Far from being a curse, its hypersensitivity to 'initial conditions' can turn out to be a blessing. By ruthlessly exploiting this property, our far-future descendants can steer the Universe along any evolutionary route they desire. All that will be necessary will be a small nudge here, a small nudge there.

Exactly what will have to be nudged where in order to make the Universe shrink down towards the Omega Point, is not clear. Figuring it out, according to Tipler, will require complex and detailed calculations.

The point is that a single, one-off redistribution of the matter of the cosmos will not be sufficient to achieve the desired Omega Point. Like a snake trying to squirm out of a jar, the chaotically fluctuating Universe will perpetually try to evolve away from the state that leads to the Omega Point – unless it is pushed back, again and again. Consequently, the occupants of this far-future universe will continually have to calculate how to correct matters. They will continually have to step in and tweak things. It will be impossible for them to relax for an instant. The price of attaining the elusive Omega Point will not only be engineering on a scarcely believable scale. It will be eternal vigilance, eternal intervention.

But steering the Universe towards the Omega Point will be only one of the problems faced by our descendants. As the Universe shrinks ever smaller, and the temperature sky-rockets towards infinity, atoms will split asunder, then the constituents of atoms, then the constituents of the constituents of atoms . . . The Universe will become a raging inferno of subatomic particles the like of which we cannot imagine with our rudimentary twenty-first-century theories of physics. It will be in the midst of this firestorm of exotic matter that our descendants will have to find a way to store the essence of their being and carry out the information processing which is synonymous with thinking.

Eventually, however, life will have to deal not with matter but the wildly fluctuating distortions of space-time. Though these will provide a tremendous source of energy, they will also create a new and unprecedented challenge. Somehow, some way, life must find a way to transfer its very essence from fiery matter into tortured space-time.

In short, if life is to survive the tumultuous journey down towards the Omega Point it will have to perpetually reinvent itself. The task will be immense but the price of failure will be oblivion. As Groucho Marx said: 'I plan to live for ever, or die trying.' That just about sums up the situation our descendants will face.

Dark Energy: a Spanner in the Works?

There is, however, a big spanner in the works. We do not appear to live in a Universe that will one day re-collapse – one whose runaway shrinkage can be corralled along the eccentric evolutionary path that leads to the Omega Point. Quite the reverse.

As pointed out before, physicists and astronomers in California and Australia discovered in 1998 that, contrary to all expectations, the expansion of the Universe appears to be speeding up. This is hard to understand because the gravity between every galaxy and every other galaxy should be acting to pull them back together again like some great cosmic web of elastic. Far from speeding up the expansion of the Universe, it ought to be braking it.

The inescapable conclusion is that gravity is not the only force orchestrating the fate of the large-scale Universe, as everyone had believed. Another, hitherto unsuspected, force must be at play. Since the Universe contains only galaxies, which dance to the tune of gravity, and the empty space between the galaxies, the mysterious force must be a property of empty space. It cannot therefore be as empty as it looks.

Instead, it must contain some kind of invisible stuff which is counteracting gravity – dark energy.

Because space contains dark energy, it is springy. The more space there is the more springiness there is. Doubling the volume of space doubles the total quantity of dark energy and doubles its repulsive effect. What this means for the Universe is that, in the beginning, when the Universe was small and there was very little space, dark energy had only a minuscule effect and gravity dominated the Universe. However, as the Universe expanded and space grew, more and more dark energy was created. Today, there is enough around that it has overwhelmed gravity.

There are two outstanding questions concerning the dark energy. The first and obvious one is – What is it? Here, physicists are utterly at sea. As already mentioned before, their best theory – quantum theory – predicts an energy for empty space which is 1 followed by 123 zeroes bigger than what is in fact observed. The second dark energy question is – Why is it gaining control of the Universe *now*? In the distant past, when there was very little space, dark energy had an entirely negligible effect. However, over the past 13.7 billion years, as space has expanded, its repulsive effect has been steadily building. Only now is it beginning to overwhelm gravity. The question is – How come we are alive at this special moment? Nobody knows the answer. But the coincidence is definitely very, very odd.

In the future, as the Universe continues to swell and dark energy grows remorselessly in importance, it will begin to drive a runaway cosmic expansion of space. Eventually, the galaxies will become infinitely isolated islands in an unimaginably vast and empty ocean of space. This is hardly what the doctor ordered if life, as Tipler believes, is to gain control of the Universe and force it down to the Omega Point. The obvious problem faced by intelligence in a universe which is growing ever faster is colonising the place. Spreading to fill every last nook and cranny of Creation is akin to completing a 100-metre race when the finishing line is receding faster and faster. But this, it turns out, is not the most serious problem posed by an accelerated universe. There is another, even more fundamental one. It has to do with the Universe's 'horizon'.

Because the Universe has been in existence for just 13.7 billion years, the only galaxies we can see are those whose light has taken less than 13.7 billion years to reach the Earth. Objects which are so far away that their light would take more than 13.7 billion years to reach us we cannot currently see. Their light is still on its way. Because of this, we can see only a limited portion of the Universe, commonly called the 'observable' Universe.*

Every year we can in principle see objects whose light has taken an extra year to travel to us. Consequently, every year, the observable Universe grows – by a light year a year – the horizon expanding outwards into the greater Universe like the surface of a swelling bubble.

A deep question is – How did so much of the Universe get to be beyond the horizon in the first place? The answer has to do with inflation, the phase of super-fast cosmic expansion that was over and done with in the first split-second of the Universe's existence. During inflation, space expanded faster than light, something which is permitted for space – the backdrop of the cosmos – but not for any material object in the Universe. Consequently, most of the Universe was stretched far from the Earth's present location – so far away that its light, though it has been travelling since the dawn of time, still has not managed to get here. The tremendous ballooning of space in effect caused the horizon of the observable Universe to shrink so that it enclosed an ever smaller portion of the greater Universe. Only when inflation had run out of steam did

---

* See Chapter 1, 'Elvis Lives'.

the horizon begin expanding again – at roughly a light year a year – gradually bringing back into the view parts of the Universe which had been driven out of sight.

The accelerated expansion of inflation fizzled out long ago and so would seem to have little relevance to the present and future Universe. Nevertheless, it does. The reason is that the Universe, after a multi-billion-year hiatus, appears once again to be embarked on a period of accelerated expansion – driven by the repulsive force of dark energy. What this means is that the horizon of the Universe will one day start shrinking again, just as it did during inflation, and the observable Universe will become an ever smaller part of the total Universe. The trouble is that a shrinking horizon has profound consequences for life in the far future and its chances of steering the Universe towards the Omega Point. It makes it difficult for life across the cosmos to coordinate its actions.

Coordination is essential in order to push the Universe towards the Omega Point. After all, matter will have to be shifted from place to place across the entire cosmos. However, our far-future descendants – or, if we are not alone, the far-future descendants of intelligent races from all the galaxies – will here face a severe difficulty. The horizon around every observer will be shrinking as the expansion of the Universe speeds up. And this means it will be impossible to see or know about a greater and greater portion of the Universe. In such circumstances, how can intelligence in different parts of the Universe possibly communicate with each other so as to coordinate their actions?

The answer is, they cannot. In a universe containing dark energy, the Omega Point option can never be engineered. Not only is it impossible in practice, it is impossible even in principle. Unless, of course, there is some way that the dark energy can be 'switched off'.

Switching off the dark energy, and with it the runaway expansion of the Universe, may seem a tall order. However, the fact that it wrested control of the fate of the Universe from gravity relatively recently in cosmic history may mean it 'switched on' recently. And, if it switched on, whatever switched it on might one day be used to switch it off. This is the belief Tipler subscribes to.

Not only does Tipler think that the dark energy can be neutralised, he actually thinks that this is inevitable. It will be an unavoidable by-product, he says, of information processing – the very characteristic that defines living things.

Tipler believes that to process more and more information life will have to use energy as efficiently as possible. The most concentrated form of energy is mass-energy. Life will therefore have to convert mass-energy into other forms of energy such as light and heat. In short, it will have to destroy mass.

No one knows for sure the origin of mass.* However, many physicists suspect that it arises from an invisible 'field' which fills all of space. As mentioned earlier, it is known as the Higgs field after the Scottish physicist Peter Higgs who proposed its existence, and it acts like treacle, impeding the motion of matter and thereby making matter difficult to budge – the property we associate with mass.

As Newton discovered, action and reaction are equal and opposite. Push against a wall and it pushes back. In the same way, Tipler believes that the destruction of mass will have a kind of 'back reaction' on the Higgs field, robbing it of its strength.

Why has this any bearing on the dark energy? Well, nobody knows what the dark energy is. Your guess is as good as mine. But Tipler identifies it with the 'cosmic repulsion which has to exist in order to cancel out the cosmic attraction of the Higgs field'. Admittedly, this is an

---

* See Chapter 8, 'Mass Medium'.

opaque statement! But, to go into any more detail here would be to obscure things yet more. Suffice to say that Tipler believes the cosmic repulsion and the cosmic attraction of the Higgs field are not in balance today – there is a small uncancelled surplus which manifests itself as the dark energy. Furthermore, he believes that the back reaction caused, as mass is destroyed across the cosmos, will gradually bring things back into balance, bit by bit depleting the dark energy. It will nullify it, gradually robbing it of its central ability – the ability to accelerate the expansion of the Universe.

How then does this solve the coordination problem? Well, the destruction of mass is simply a by-product of the information processing that creatures will be doing naturally all over the Universe. There will be no need for them to coordinate their actions. There will be no need for them to know about each other's existence. The destruction of mass will inevitably occur everywhere across the Universe. And, with it, the dark energy will be neutralised.

Of course, once the dark energy is neutralised the horizon around each observer will cease to shrink and begin to expand. Eventually, everyone in the Universe will be able to see everyone else. There will be no barrier to cosmic coordination. There will be no obstacle to a universe-wide engineering project. The minor hiccup of the dark energy having been dealt with, it will be possible to steer the Universe all the way down towards the elusive Omega Point.

The Omega Point

As the Universe shrinks closer and closer to the Omega Point, the amount of information processing that can be done will of course sky-rocket without limit. To perpetually keep on top of things our descendants will have to adapt to ever more extreme conditions of density and temperature, which will involve them transferring their minds – their essential software – into hardware composed of ever more exotic subatomic particles. Provided that our descendants can find a way to keep on doing this, however, stretching before them like a never-ending road, will be a future of infinite promise, infinite subjective time.

Speculating on what super-intelligent life might do with all this time on its hands is a risky business, to say the least. Nevertheless, as Dyson has pointed out, if we wish to contemplate the long-term fate of the Universe, we have no choice but to grapple with the question of the values and purposes of intelligent life. And, on this subject, Tipler believes his guesses are as good as anyone else's.

One thing a super-intelligence might want to do, says Tipler, is harness the phenomenal information-processing power close to the Omega Point universe to create simulations of real life. With so much computing power available, such simulations could be rendered with such extraordinary fidelity that they would be indistinguishable from the real thing.* And this, according to Tipler, opens up a fascinating possibility. It all hinges on another remarkable property of the Omega Point universe.

The Omega Point is not simply a point in space and time with the capacity to carry out an infinite amount of information processing. It is a point in space and time onto which converge light rays from the entire past history of the Universe. This is an extraordinary and significant fact. Just think. Before converging on the Omega Point, those light rays will have bounced off stars and galaxies and planets throughout the length

---

* The perfect simulation of reality is a possibility because of the ability of a computer – otherwise known as a Universal Turing Machine – to mimic the operation of any conceivable machine (see Chapter 6, 'God's Number'). One such machine is the world around us.

and breadth of the past Universe. Consequently, they will carry with them information about the location and arrangement not only of every chunk of inert matter but also of every living creature that ever existed. With such information, it will be possible to create the ultimate computer simulation – a simulation of everything that has ever existed in the Universe. 'What we're talking about, in effect, is "resurrecting" each and every one of us – in fact, every creature who has ever lived,' says Tipler.

Once again, we come up against the sticky problem of anticipating the motivations of a super-intelligence – a super-intelligence moreover whose thought processes will likely be as far beyond ours as ours are beyond those of the lowliest bacterium. Tipler, however, is undaunted. 'The reason our far-future descendants will want to run the ultimate simulation will be simply to find out all they can about their past,' he says. 'Just like us, they will have burning desire to know exactly where they came from.'

The ultimate simulation is not, however, without its problems – even for an intelligence that possesses near-infinite computing resources. The reason has to do with the all-important light rays funnelling down to the Omega Point. Although the information they carry about the past history of the Universe is available in principle, extracting it will in practice be exceedingly difficult. The torrent of light raining down onto the Omega Point will be tremendous. Disentangling anything useful from it will be like picking out the individual voices from the roar of a football crowd stadium – only hugely, hugely harder.

All is not lost, though. Even if the past cannot be deduced in its entirety from the light rays converging on the Omega Point, Tipler sees another, ingenious, way to resurrect every creature that has ever lived. It all depends on a quantum theory – our description of the microscopic world of atoms and their constituents – and one 'interpretation' of quantum theory in particular.

As pointed out before, quantum theory predicts something bizarre about the world: an atom can be in many places at once. This is not some weird theoretical prediction. In experiments, it is in fact possible to observe an atom being in two places at once – or, more accurately, the consequences of an atom being in many places at once. But if an atom can be in many places at once, how is it that, when atoms come together to make big things like chairs, tables and people, these big things cannot be in many places at once? According to the standard explanation, some process intervenes between the scale of atoms and scale of tables to force them to behave themselves and stop being in many places at once. But there is another, arguably simpler, explanation, which is increasingly favoured by physicists. And that is that nothing intervenes. In other words, the small-scale world is exactly the same as the large-scale world. And, just as an atom can be in many places at once, so too can a table, a chair and a person.

Why then do we never see a person in two places at once – for instance, going through two adjacent doors at the same time? The answer is because the two possibilities happen in separate, or parallel, realities. This is the Many Worlds idea that there are an infinity of parallel realities in which all possible histories are played out.* In some realities there are versions of you living similar lives; in other realities versions living very different lives; in yet other realities there are no versions of you at all because you were never born.

Tipler – no prizes for guessing – is a strong proponent of the Many Worlds idea. Consequently, he believes that, even if the creatures at the end of time cannot extract the information necessary to re-create the past Universe from the light funnelling down towards the Omega Point, there will be another option open to them. They can simulate all the

* See Chapters 1 and 4, 'Elvis Lives' and 'Keeping It Real'.

possible realities of the Many Worlds that lead up to the Omega Point. Among this uncountable profusion of past histories of the Universe, inevitably they will find the history followed by their ancestors.

Of course, simulating all possible histories of the Universe requires mind-bogglingly more computer resources than simulating a solitary history. But this is no problem, according to Tipler. If, at any time, our descendants have insufficient computing resources to carry out the necessary mega-simulation, they will merely have to wait a little longer. With the Universe's ability to process information sky-rocketing so rapidly, pretty soon they will find themselves in possession of the necessary computing resources, no matter how great those resources might be.

Inevitably, among all the possible realities leading up to the Omega Point, will be the reality that contains you and me. 'One way or another, we are going to find ourselves resurrected in a computer simulation at the end of time,' says Tipler.

Wait. How do we know that we are not in a computer simulation at this very moment? How do we know, as is remarked in the film *Breathless*, that we are not 'the dead on vacation'? Outrageous as it seems, some people think it is a serious possibility that we are currently in a giant computer simulation, though not necessarily the Omega Point simulation envisaged by Tipler.

Are We in a Computer Simulation?

The argument is based on a simple premise: at some time in the future, if technological progress continues unabated, it will be possible to build computers powerful enough to mimic human consciousness. If this premise is accepted, says the philosopher Nick Bostrom of Oxford University, there are only three possible future scenarios.*

In the first scenario, some kind of global catastrophe wipes out the human race before it can build the necessary super-computers. If this is to be our fate, we can at least console ourselves with the thought that the reality we are currently experiencing is definitely real. In the second future scenario, we develop the necessary super-computers but have no interest whatsoever in running simulations. Though not impossible to believe, this may not be too likely given our proven habit of doing things, such as trigger nuclear chain reactions or clone human beings, 'just to see what happens'. That just leaves the third scenario.

In this final future, we will not only simulate conscious beings but also the universes in which they will live. If this scenario is the true one, then the likelihood is that the simulations have already been done and we are at this moment living in an artificial reality!

It is always possible, of course, that we are living in the pre-simulation, 'real' world. However, Bostrom points out that the future stretching before us is vast – time enough to run a large number of simulations. With the overwhelming majority of simulations being artificial, what is the chance that we should find ourselves in the only simulation that is truly real? Bostrom thinks the answer is clear. No chance. We are living in someone's computer simulation with 100 per cent certainty.

But, if we are in a computer simulation, how can we ever tell? Clearly, if the rendering of reality is near-perfect – as Tipler believes is possible arbitrarily close to the Omega Point – it will be extremely difficult to distinguish pseudo-reality from reality. Nevertheless, even the best computer simulations have their flaws.

* Bostrom's paper 'The Simulation Argument' from the journal *Philosophical Quarterly* (vol. 53, no. 211, pp. 243–255, 2003) is at: *http://www.simulation-argument.com/*

In any simulation, it is not possible to record to an unlimited degree of accuracy the position, speed, and so on, of every particle of matter. In practice, such numbers have to be truncated to a finite number of digits. Inevitably, this leads to tiny errors creeping into the simulation. And, with each new round of number-crunching, which is necessary to maintain the simulation, such tiny errors will get magnified. Eventually, like the butterfly whose flapping wings spawn a hurricane, they will have a significant and noticeable effect.

Putting an error-ridden simulation back on track requires reaching into the computer on which the simulation is running and resetting the numbers. If we are indeed living in simulation then, from time to time, such computational course corrections are unavoidable. How would they manifest themselves? Well, the most likely way is as sudden changes in the laws of physics – an abrupt jump in the strength of gravity, perhaps, or in the electrical charge carried by the electron.

Remarkably, just such a change appears to have happened in our Universe many billions of years ago. The change has revealed itself in the light of distant quasars, the super-bright cores of newborn galaxies.*

Their prodigious brightness means that they can be seen at enormous distances. And, since their light has taken so long to travel across space to us, we see quasars as they were when the Universe was in its youth, many billions of years ago.

From the light coming from quasars astronomers can deduce the elements they contain. This is because the atoms of a particular element

---

* A quasar typically pumps out the light energy of 100 normal galaxies – that is ten million million suns – from a tiny region smaller than our Solar System. Such objects are believed to be powered by rapidly spinning black holes, up to ten billion times the mass of the Sun. In a quasar, matter swirls down through an 'accretion disc' onto a 'supermassive' black hole. As it does so, it is heated to millions of degrees by internal friction. It is the light from this super-hot accretion disc that we see as the burning beacon of the quasar.

emit light at wavelengths which are as unique to that element as a fingerprint is to an individual person. If astronomers detect the light at the wavelength's characteristic of calcium, for instance, they know for sure that the quasar contains calcium.

In 1998, a team of astronomers from the University of New South Wales in Australia and the University of Sussex in Britain reported that it had found subtle differences between the fingerprint of atoms of a particular element on Earth today and atoms of the very same element in quasars about ten billion years in the past.

If correct – and no one has yet managed to show that the observation is in error – this is an extraordinary discovery. The light given out by atoms is determined by a number known as the 'fine-structure constant'. This orchestrates the delicate interplay between light and matter. In doing so, it determines how tightly normal matter, including the matter in our bodies, is glued together. If ten billion years ago the light given out by atoms was different, it can only mean that the fine-structure constant must also have been different. In fact, the observations show it was several parts in 100,000 smaller than today.

So, did the simulation we live in go awry at some time in the past ten billion years and have to be corrected by whoever or whatever is running it? And, if it did, can we expect other corrections? Will the speed of light suddenly jump one day? Or the strength of the gravity pinning our feet to the ground? Or will we look up one day and see the kind of sight described by Arthur C. Clarke in his 1953 short story, 'The Nine Billion Names of God'?*

'Look,' whispered Chuck, and George lifted his eyes to heaven.

* Collected in *The Other Side of the Sky* by Arthur C. Clarke (VGSF, London, 1987).

(There is a last time for everything.) Overhead, without any fuss, the stars were going out.

Tipler has no particular view on whether or not we are currently living in such a generic computer simulation. However, he has a strong view on whether or not we are living in the ultimate simulation – the one near the Omega Point at the end of time. 'It is extremely unlikely,' he says.

Tipler's reasoning is interesting. 'Our world is so imperfect,' he says. 'Why would an intelligence with pretty much unlimited computing power go to the trouble of simulating such a flawed reality?'

Of course, Tipler is once again in the position of guessing the motivations of an intelligence vastly in advance of our own. Though many would say this is impossible, Tipler believes he can anticipate at least some of the things such a super-intelligence might want to do with its superpowers. To support his belief, he points to some remarkable features of the Omega Point – features which give any intelligence in its vicinity extraordinary, and deeply suggestive, abilities.

God and the Omega Point

As discussed earlier, a super-intelligence close to the Omega Point will be able to carry out an unlimited amount of information processing. It will be able to resurrect everyone who has ever lived and give them eternal life. 'With this kind of power to create and manipulate reality, it will be omnipotent,' says Tipler.

But this is not all.

As also mentioned earlier, all the myriad light rays from the entire past of the Universe converge on the Omega Point. Since these light rays carry with them knowledge about everything and everyone that ever existed, any super-intelligence near the Omega Point will be all-seeing. 'They will be omniscient,' says Tipler.

See where this is going? Tipler – controversially – identifies the Omega Point with God! 'People talk of God as the creator of life,' he says. 'But maybe the purpose of life is to create God.'

Tipler, by making this outrageous claim, is stretching physics way beyond its widely accepted boundaries and striking deep into the territory of theology. It goes without saying that the majority of physicists think he has stretched physics far beyond its breaking point. They think it is important to draw a line in the sand between what is science – the possibility of life surviving for ever at the Omega Point – and what is theological speculation.

Tipler, however, makes no apology for his claim. He points out that the stated aim of physics is to describe the Universe in its entirety. 'If it is to succeed in this task, clearly it must also describe any Supreme Being living in the Universe,' he says. 'It therefore follows that theology must eventually be shown to be a branch of physics.'

If Tipler's identification of the Omega Point with God is taken seriously, then, bizarrely, there is a connection between God and the matter content of the Universe. After all, the only way the Universe can shrink – either down to a standard Big Crunch or to a non-standard Omega Point – is if it contains sufficient gravitating mass to slow, then slam into reverse, its headlong expansion. Having the right mass is the key. 'The existence of God depends on the amount of matter in the Universe,' says Tipler.

Of course, there is a subtle difference between God, in the widely accepted sense of the word, and a location in space-time with many of the attributes of God, which is Tipler's Omega Point. Nevertheless, any intelligence inhabiting the tortured space-time close to the Omega

Point will, by virtue of its omniscience and omnipotence, have God-like powers. It will, in effect, be God.

So now we come back to Tipler's conviction that, although we will all be resurrected in a computer simulation at the end of time, we are not actually living in that simulation at present. What makes him so sure of this is his belief that any God-like entity which is both omnipotent and omniscient would likely be infinitely benevolent as well. In other words, they would be highly unlikely to simulate a reality which contained poverty, unhappiness, war and misery – one, that is, that closely corresponds to the world we find ourselves in.

According to Tipler, bad things simply will not happen in the simulation at the end of time. Unhappiness will not exist. Once again, Tipler cannot resist using a highly emotive word to describe what it will be like. 'It'll be Heaven,' he says.

Your destiny, he claims, is to die, then wake up in a computer simulation that is indistinguishable from the Judaeo-Christian Heaven. What will it feel like? Well, dead people, by definition, perceive nothing. So, although trillions upon trillions of years will separate your death on Earth from your rebirth near the Omega Point, you will be totally unaware of this yawning chasm of time. One moment you will close your eyes and the next you will open them at the end of time. From your point of view, your resurrection will be instantaneous.

It is impossible not to marvel at the temerity of Tipler. In the Omega Point theory, he has taken science and stretched it to within a hair's breadth of breaking point. If Tipler is right, none of us will die. We all meet up with our family and friends at the great party at the end of time. 'If anyone has lost a loved one, or is afraid of death,' writes Tipler, 'modern physics says: "Be comforted, you and they shall live again".' Whether or not you believe him, there is no doubt that Tipler has journeyed to the far shores of modern physics.

Afterword
The Ultimate Question

We've come a long way. But there is still a long way to go. In the late nineteenth century, physicists confidently claimed they had discovered pretty much all there was to be discovered. That was before the twin bombshells of quantum theory and relativity blew the edifice of nineteenth-century, 'classical' physics sky high. A hundred years later, physicists such as Stephen Hawking − not overly quick to learn their lesson − were predicting we would have a 'theory of everything' within a decade or so. That was before the unexpected discovery in 1998 that you, me, the stars and the galaxies account for a mere 4 per cent of the mass of the Universe. The cosmos instead is dominated by mysterious invisible stuff with repulsive gravity − dark energy − not to mention invisible stuff with ordinary gravity − dark matter. And, if any further proof is needed that the theory of everything might be farther away than the optimists think, there is the small matter that quantum theory − our very best physical theory − overestimates the energy density of dark matter by a factor of 1 followed by 123 zeroes!

Perhaps some novel idea like string theory will tie together quantum theory and general relativity in a neat ten-dimensional package. In some quarters there is great hope that this will happen – though perhaps not soon. In other quarters there is scepticism about a 'bandwagon' theory which makes few, if any, predictions that are testable in the short-term. They point out that the theory is undoubtedly beautiful but that there are many equally beautiful ideas which the Creator in his wisdom has chosen not to implement.

Maybe string theory will be the saviour of modern physics. Then again, perhaps it won't. One thing is in no doubt, however. Scientifically, we live in extremely interesting times. We can ask questions our forebears couldn't even formulate and expect to get answers well within a lifetime. Some of the questions we can ask I have addressed in this book. But of course many are left:

* What is dark energy?
* What is dark matter?
* What is space?
* What will be the ultimate fate of the Universe?
* Why is the speed of light the ultimate cosmic speed limit?
* What is consciousness?
* Does life have any special role to play in the Universe?

Then, of course, there is that most puzzling question of all: Why has the Universe given rise to matter that contemplates its surroundings and asks 'why?' Now that is a question and a half. I think I can feel another book coming on . . .

Glossary

Absolute Zero Lowest temperature attainable. As a body is cooled, its atoms move more and more sluggishly. At absolute zero, equivalent to -273.15 on the Celsius scale, they cease to move altogether. (In fact, this is not entirely true since the Heisenberg uncertainty principle produces a residual jitter even at absolute zero.)

Acceleration The rate of change of velocity. Since velocity is defined as having both a magnitude (speed) and a direction, a body accelerates whenever it changes its speed, its direction or both. For instance, a car pulling away from traffic lights accelerates as does the Moon circling the Earth at pretty much constant speed.

Accretion disc CD-shaped disc of in-swirling matter that forms around a strong source of gravity such as a black hole. Since gravity weakens with distance from its source, matter in the outer portion of the disc orbits more slowly than in the inner portion. Friction between regions where matter is travelling at different speeds heats the disc to millions of degrees. Quasars are thought to owe their prodigious brightness to ferociously hot accretion discs surrounding 'supermassive' black holes.

Algorithmic Information Theory A field of mathematics in which the complexity of a number is defined as the length of the shortest computer program capable of generating the number. Since it is actually never possible to know for certain that you have found the shortest program – determining whether you have is an uncomputable problem – the whole field is founded on paradox and uncertainty.

Alpha Centauri The nearest star system to the Sun. It consists of three stars and is 4.3 light years distant.

Angular momentum A measure of how difficult it is to stop a rotating body.

Angular momentum, conservation of The edict that, in the absence of any turning force – torque – angular momentum can never be created or destroyed.

Annihilation The mutual destruction of a subatomic particle and its antiparticle when they run into each other. In the process, their mass-energy is converted into radiant energy – gamma rays.

Anthropic principle The idea that the Universe is the way it is because, if it was not, we would not be here to notice it! In other words, the fact of our existence is an important scientific observation.

Antimatter A large accumulation of antiparticles. Anti-protons, anti-neutrons and positrons can in fact come together to make anti-atoms. And in principle there is nothing to rule out the possibility of anti-stars, anti-planets and anti-life. One of the greatest mysteries of physics is why we appear to live in a Universe made solely of matter when the laws of physics seem to predict a 50/50 mix of matter and antimatter.

Antiparticle Every subatomic particle has an associated antiparticle with opposite properties such as electrical charge. For instance, the negatively charged electron is twinned with a positively charged antiparticle known as the positron. When a particle and its antiparticle meet, they self-destruct, or 'annihilate', in a flash of high-energy light, or gamma rays.

Atom The building block of all normal matter. An atom consists of a nucleus orbited by a cloud of electrons. The positive charge of the nucleus is exactly balanced by the negative charge of the electrons. An atom is about a ten-millionth of a millimetre across.

Atomic energy See Nuclear energy.

Atomic nucleus The tight cluster of protons and neutrons (a single proton in the case of hydrogen) at the centre of an atom. The nucleus contains more than 99.9 per cent of the mass of an atom.

Axiom A self-evident truth. Each field of mathematics is built on a set of axioms from which theorems are deduced by the application of logic.

Big Bang The titanic explosion in which the Universe is thought to have been born 13.7 billion years ago. 'Explosion' is actually a misnomer since the Big Bang happened everywhere at once and there was no pre-existing void into which the Universe erupted. Space and time and energy all came into being in the Big Bang.

Big Bang theory The idea that the Universe began in a super-dense, super-hot state 13.7 billion years ago and has been expanding and cooling ever since.

Big Crunch If there is enough matter in the Universe, its gravity will one day halt and reverse the Universe's expansion so that it shrinks down to a Big Crunch. This is a sort of mirror-image of the Big Bang.

Binary A way of representing numbers. In everyday life we use decimal, or base 10. The right-hand digit represents the 1s, the next digit the 10s, the next the 10 × 10s, and so on. For instance, 9217 means $7 + 1 \times 10 + 2 \times (10 \times 10) + 9 \times (10 \times 10 \times 10)$. In binary, or base 2, the right-hand digit represents the 1s, the next digit the 2s, the next the 2×2s, and so on. So, for instance, 1101 means $1 + 0 \times 2 + 1 \times (2 \times 2) + 1 \times (2 \times 2 \times 2)$, which in decimal is 13. Binary is particularly useful for computers which are composed of 'transistors' – devices with two states.

Black hole The grossly warped space-time left behind when a massive body's gravity causes it to shrink down to a point. Nothing, not even light, can escape, hence a black hole's blackness. The Universe appears to contain at least two distinct types of black hole – stellar-sized black holes, formed when very massive stars can no longer generate internal heat to counterbalance the gravity trying to crush them, and 'supermassive' black holes. Most galaxies appear to have a super-massive black hole in their heart. They range from millions of times the mass of the Sun in our Milky Way to billions of solar masses in the powerful quasars.

Brane A lower-dimensional 'island' in a higher-dimensional space. In superstring theory, for instance, space has ten dimensions, which has led to the suggestion that our Universe is a four-dimensional brane, or four-brane.

Butterfly effect The fact that in a chaotic system – which is infinitely sensitive to initial conditions – an arbitrarily small change in the initial state of the system can in time produce a dramatic effect. For instance, in the case of the global weather system, the beat of a butterfly's wings in China can eventually trigger a hurricane in the Caribbean.

Carrier wave A periodic wave on which an information-carrying signal is superimposed. In the case of a radio wave, for instance, the frequency or the amplitude is continuously varied, or 'modulated'.

Cellular automaton A simple computer program that takes an input – a pattern of coloured cells – and applies a simple rule to the input to produce an output – another pattern of coloured cells. The key thing is that the output is fed back in as the next input rather like a snake swallowing its own tail. In the case of a two-colour, adjacent cell, one-dimensional cellular automaton, the input is the pattern of black-and-white cells on one line and the output is the pattern of cells on the next line. Whether a cell in the second line is black or white depends on a rule applied to its two nearest neighbours in the first line. The rule might say, for instance: 'If a particular cell in the first line has a black square on either side of it, it should turn black in the second line.'

Cellular automaton rule 110 One of the 256 possible rules for a two-colour, adjacent cell, one-dimensional cellular automaton. The rule is special because it generates unending complexity – a pattern that never repeats itself.

CERN The Organisation Européenne pour la Recherche Nucléaire (European Organization for Nuclear Research) – the laboratory for particle physics near Geneva, Switzerland. The acronym originates from the council that set it up, the Conseil Européen pour la Recherche Nucléaire (European Council for Nuclear Research).

Chaos A system which is infinitely sensitive to initial conditions – for instance, the global weather system. This makes it unpredictable not in principle but in practice – because it is never possible to know the starting conditions precisely enough to make a long-term prediction.

Classical physics Non-quantum physics. In effect, all physics before 1900 when the German physicist Max Planck first proposed that energy might come in discrete chunks, or 'quanta'. Einstein was the first to realise that this idea was totally incompatible with all physics that had gone before.

Cloud chamber A sealed box with a window in which ultra-pure water vapour is cooled well below the temperature at which it normally condenses to form water droplets. The physicist Charles Wilson invented it to investigate how clouds form but soon realised it could be used to reveal the tracks of subatomic particles such as electrons. Water droplets formed like tiny beads around the ions created by the subatomic particles.

Comet Snowy body – usually only a few kilometres across – that orbits a star. Most comets orbit the Sun beyond the outermost planets in an enormous cloud known as the Oort Cloud. Like asteroids, comets are builders' rubble left over from the formation of the planets.

Compton wavelength The size an electron appears to have in experiments when photons rebound, or 'scatter', off it.

Computational irreducibility The property of a computer program whose output can be deduced from its input only by running the program. There is no shortcut.

Computational reducibility The property of a computer program whose output can be deduced from its input in less time than it takes to run the program. There is a shortcut.

Computer program A series of instructions which is applied to an input to create an output.

Computer program, recursive A computer program whose output is fed back in as its next input like a snake swallowing its tail.

Conservation law Law of physics that expresses the fact that a quantity can never change. For instance, the conservation of energy states that energy can never be created or destroyed, only converted from one form to another. For example, the chemical energy of petrol can be converted into the energy of motion of a car.

Copenhagen Interpretation of quantum theory An attempt to explain why denizens of the microscopic world such as an electron exhibit bizarre 'quantum' behaviour – for instance, being in two places at once – while those of the everyday world do not. According to the Copenhagen, the act of 'observing' an electron somehow forces it to behave and plump for being in one place at once. Since the act of observation is not defined, the whole interpretation is open to interpretation. Most commonly, physicists talk of a big, or 'macroscopic', object observing a microscopic object and killing its quantumness. Consequently, the Copenhagen Interpretation splits the world into two – small things that obey quantum theory – and big things which do not.

Copernican principle The idea there is nothing special about our position in the Universe, either in space or in time. This is a generalised version of Copernicus's recognition that the Earth is not in a special position at the centre of the Solar System but is just another planet circling the Sun.

Cosmic Background Explorer Satellite (COBE) Satellite launched in 1989 to 'map' the temperature of the cosmic background radiation – the 'afterglow' of the Big Bang fireball – across the sky. COBE found slight variations in the average temperature of the radiation, which were created by matter beginning to clump 450,000 years after the birth of the Universe. The clumps were the 'seeds' of giant superclusters of galaxies in today's Universe.

Cosmic Background Radiation The 'afterglow' of the Big Bang fireball. Incredibly, it still permeates all of space 13.7 billion years after the event, a tepid radiation corresponding to a temperature of -270°C.

Cosmic rays High-speed atomic nuclei, mostly protons, from space. Low-energy ones come from the Sun, high-energy ones probably come from supernovae. The origin of ultra-high-energy cosmic rays, particles millions of times more energetic than anything we can currently produce on Earth, is one of the great unsolved puzzles of astronomy.

Cosmology The ultimate science. The science whose subject matter is the origin, evolution and fate of the entire Universe.

Cosmos Another word for Universe.

Coupling constant A number which is a measure of the strength of one of nature's fundamental forces. For instance, the fine-structure constant governs the interaction between light (photons) and matter. Since this interaction is the basis of the electromagnetic force, it governs the strength of this force.

Dark energy Mysterious 'material' with repulsive gravity. Discovered unexpectedly in 1998, it is invisible, fills all of space and appears to be pushing apart the galaxies and so speeding up the expansion of the Universe. It accounts for about 73 per cent of the mass-energy of the Universe and nobody has much of a clue what it is! In fact, quantum theory predicts that it should have an energy density 10^{123} times bigger than it actually does.

Dark matter Matter in the Universe which gives out no detectable light. Astronomers know it exists because the gravity of the invisible stuff bends the paths of visible stars and galaxies as they fly through space. There is at least six times as much dark matter in the Universe as visible matter. The identity of the dark matter is the outstanding problem of astronomy.

Davies–Unruh effect The heat radiation a person accelerating through the quantum vacuum sees coming from in front of them. Its origin is in the virtual particles which are constantly popping in and out of existence in the quantum vacuum. From the point of view of an accelerated observer, they appear much like the heat from a furnace and have a temperature dependent on the person's acceleration.

Decoherence The loss of the weird quantum nature of a microscopic particle when it makes a record on some kind of detector – for instance, the eye. All weird quantum behaviour ultimately stems from interference between the individual components of a wave representing a particle. If those waves do not overlap – if they are not coherent – there can be no interference between the waves and no weird quantum behaviour. This is the case when a particle leaves a record on the large number of atoms of a detector like the eye. Wave overlap, or coherence, is irretrievably lost. Hence the term decoherence.

Dimension An independent direction in space-time. The familiar world around us has three space dimensions (left–right, forward–backward, up–down) and one of time (past–future). Superstring theory requires the Universe to have six extra space dimensions. These differ radically from the other dimensions because they are rolled up very small.

Density The mass of an object divided by its volume. Air has a low density and iron has a high density.

Double-slit experiment Experiment in which subatomic particles are shot at a screen with two closely spaced, parallel slits cut in it. On the far side of the screen, the particles mingle, or 'interfere', with each other to produce a characteristic 'interference pattern' on a second screen. The bizarre thing is that the pattern forms even if the particles are shot at the slits one at a time, with long gaps between – in other words, when there is no possibility of them mingling with each other. This result, claimed the American physicist Richard Feynman, highlighted the 'central mystery' of quantum theory.

DNA Deoxyribonucleic acid, the ultimate store of genetic information for all cells.

Einstein's theory of gravity See Relativity, general theory of.

Electric charge A property of microscopic particles which comes in two types – positive and negative. Electrons, for instance, carry a negative charge and protons a positive charge. Particles with the same charge repel each other while particles with opposite charges attract.

Electric charge, conservation of The edict that electric charge can never be created or destroyed.

Electric field The field of force which surrounds an electric charge.

Electromagnetic field The field which underpins the electromagnetic force.

Electromagnetic force One of the four fundamental forces of nature and the one responsible for gluing together ordinary matter such as the atoms in our bodies.

Electromagnetic wave A wave that consists of an electric field which periodically grows and dies alternating with a magnetic field which periodically dies and grows. An electromagnetic wave is generated by a vibrating electric charge and travels through space at the speed of light. In fact, it is light.

Electron Negatively charged subatomic particle typically found orbiting the nucleus of an atom. As far as anyone can tell, it is a truly elementary particle, incapable of being subdivided.

Electron–positron pair Exactly what it says. When an electron is created, it is always created alongside its antiparticle, the positron.

Energy A quantity which is almost impossible to define! Energy can never be created or destroyed, only converted from one form to another. Among the many familiar forms are heat energy, energy of motion, electrical energy, sound energy, and so on.

Energy, conservation of The principle that energy can never be created or destroyed, only converted from one form to another.

Energy, heat One of the myriad forms of energy. Since heat is actually microscopic disorder – for instance, the chaotic motion of gas atoms, flying hither and thither, in a gas – it is the lowliest form of energy, inasmuch as, eventually, all forms of energy get degraded into heat-energy, the ultimate slag of the Universe.

Energy of motion The energy a body possesses by virtue of the fact it is moving.

Expanding Universe The fleeing of the galaxies from each other in the aftermath of the Big Bang.

Event horizon The imaginary one-way 'membrane' that surrounds a black hole. Anything that falls through – whether matter or light – can never get out of the hole again.

Field An entity which fills all of space, assigning to each point a unique property. In the case of the gravitational field, for instance, the property is the direction and strength of the force of gravity. In modern physics, fields are considered more fundamental than forces.

Field, colour The field which gives rise to the 'strong nuclear', or colour, force between quarks.

Field, Higgs Hypothetical field which fills the vacuum like cosmic treacle. By sticking to particles, it bestows mass on them.

Fine-structure constant The number – or coupling constant – which determines the strength of the interaction between light and matter. In everyday language, it determines the strength of the electromagnetic force which glues together the atom in our bodies.

Force, centrifugal A fictitious force which we invent to explain why we appear to be flung outwards when we, for instance, round a bend in a car. In fact, there is no such force and we are not flung outwards. We are merely continuing to travel in a straight line under our own inertia and it is the body of the car that is changing its motion, not us.

Force, colour One of the four fundamental forces of nature and the one responsible for gluing together the quarks inside the nuclei of atoms.

Force, electroweak A force which exists at high energies and therefore existed in the early moments of the Universe's existence. The electromagnetic force and the weak nuclear force are both facets of this unified force.

Force, fictitious A force we invent to explain our motion when we refuse to recognise the reality – that we are in fact moving solely under our own inertia and that it is our surroundings that are changing their motion, not us.

Force, fundamental One of the four basic forces which are believed to underlie all phenomena. The four forces are gravitational force, electromagnetic force, strong force and weak force. The strong suspicion among physicists is that these forces are in fact merely facets of a single superforce. In fact, experiments have already shown the electromagnetic and weak forces to be different sides of the same coin.

Force, GUT A hypothetical force predicted by 'Grand Unified Theories' to exist at high energy, and therefore to have existed in the very early Universe. Today's electromagnetic, weak nuclear and strong nuclear forces are mere facets of this unified force.

Force-carrying particle Microscopic conveyor of a force. Forces arise through the continual exchange of such particles in much the same way that the continual exchange of a tennis ball between tennis players results in a force between them.

Fourier analysis A mathematical technique for picking out patterns which repeat in a waveform or signal.

Frequency How fast a wave oscillates up and down. Frequency is measured in Hertz (Hz), where 1 Hz is 1 oscillation per second.

Frequency band A range of frequencies.

Fundamental particle One of the basic building blocks of all matter. Currently, physicists believe there are 6 different quarks and 6 different leptons, making a total of 12 truly fundamental particles. The hope is that the quarks will turn out to be merely different faces of the leptons.

Galaxy One of the basic building blocks of the Universe. Galaxies are great islands of stars. Our own island, the Milky Way, is spiral in shape and contains about 200 hundred thousand million stars.

Gamma ray The highest-energy light, or electromagnetic wave.

General theory of relativity See Relativity, general theory.

Geodesic The shortest path between two points in warped, or curved, space.

Global positioning system (GPS) System of satellites, each carrying a clock and broadcasting a precise timing signal. A receiver on the ground determines its position by comparing the different amounts of time the various signals take to reach it.

Gluon Force-carrying particle of the strong nuclear force.

Gödel's incompleteness theorem The proof that, for any set of axioms, there exist theorems which cannot be deduced from the axioms – undecidable theorems, which are true but can never be proved to be true.

Grand unified theories (GUTs) Theories which attempt to show that, at high energies, and consequently early on in the Big Bang, the three electromagnetic, weak nuclear and strong nuclear forces were combined into a singly unified force.

Gravitational force The weakest of the four fundamental forces of nature. Gravity is approximately described by Newton's universal law of gravity but more accurately described by Einstein's theory of gravity – the general theory of relativity. General relativity breaks down at the singularity at the heart of a black hole and the singularity at the birth of the Universe. Physicists are currently looking for a better description of gravity. The theory, already dubbed quantum gravity, will explain gravity in terms of the exchange of particles called gravitons.

Gravitational wave A ripple spreading out through space-time.

Gravitational waves are generated by violent motions of mass such as the merger of black holes. Because they are weak, they have not been detected directly yet.

Gravity See Gravitational force.

Gravity, repulsive The gravity of a material with a very large negative pressure – strictly speaking, a pressure which is less than $-\frac{1}{3}$ the material's energy density. The vacuum is thought to have possessed this property at the beginning of the Universe and to have therefore 'inflated'. Today's Universe also seems to be dominated by such material, dubbed 'dark energy', which is remorselessly driving the galaxies apart.

Halting problem The remarkably simple problem that the Mathematician Alan Turing discovered was uncomputable by any conceivable computer. Given a computer program, is it possible to tell ahead of actually running the program whether it will eventually halt? The answer is no.

Hardware The unchangeable components of a computer – for instance, its transistors and disc drives.

Hawking radiation The heat radiation which is generated near the event horizon of a black hole. A consequence of quantum theory, it arises because pairs of virtual particles and their antiparticles are continually popping in and out of existence in the vacuum, as permitted by the Heisenberg uncertainty principle. Near the horizon of a black hole, however, it is possible for one particle of a pair to fall into the hole. The left-behind particle, with no antiparticle to annihilate with, is boosted from a virtual particle to a real particle. Such particles stream away from a black hole – though admittedly the effect is small for a stellar black hole – as radiation with a characteristic temperature.

Heat See Energy, heat.

Heat death The sorry fate of a universe that expands for ever. More and more of the energy of the Universe ends up as heat – microscopic disorder – characterised by a single temperature. Since all cosmic activity is ultimately driven by temperature differences, the Universe dies a death.

Heisenberg uncertainty principle A principle of quantum theory that there are pairs of quantities such as a particle's location and speed that cannot simultaneously be known with absolute precision. The uncertainty principle puts a limit on how well the product of such a pair of quantities can be known. In practice, this means that if the speed of a particle is known precisely, it is impossible to have any idea where the particle is. Conversely, if the location is known with certainty, the particle's speed is unknown. By limiting what we can know, the Heisenberg uncertainty principle imposes a 'fuzziness' on nature. If we look too closely, everything blurs like a newspaper picture dissolving into meaningless dots.

Higgs field See Field, Higgs.

Higgs particle See Particle, Higgs.

Horizon problem The problem that far-flung parts of the Universe which could never have been in contact with each other, even in the Big Bang, have almost identical properties such as density and temperature. Technically, they were always beyond one another's horizons. The theory of inflation provides a way for such regions to have been in contact in the Big Bang and so can potentially solve the horizon problem.

Horizon of Universe The Universe has a horizon much like the horizon that surrounds a ship at sea. The reason for the Universe's horizon is that light has a finite speed and the Universe has been in existence for only a finite time. This means that we only see objects whose light has had time to reach us since the Big Bang. The observable Universe is therefore like a bubble centred on the Earth, with the horizon being the surface of the bubble. Every day the horizon expands outwards and new things become visible, just like ships coming over the horizon at sea.

Hydrogen burning The fusion of hydrogen into helium accompanied by the liberation of large quantities of nuclear binding energy. This is the power source of the Sun and most stars.

Inertia The tendency for a massive body, once set in motion, to keep on moving – at constant speed in a straight line in unwarped space and along a geodesic in warped space. Nobody knows the origin of inertia.

Inflation, eternal A generic property of inflation. Although the inflationary, or false, vacuum continually decays into bubbles of normal vacuum – creating Big Bang universes – the false vacuum grows in volume at a faster rate than it is lost. Consequently, inflation, once begun, is unstoppable.

Inflation, theory of Idea that in the first split-second of the creation, the Universe underwent a fantastically fast expansion. In a sense inflation preceded the conventional Big Bang explosion. If the Big Bang is likened to the explosion of a hand grenade, inflation was like the explosion of an H-bomb. Inflation can solve some problems with the Big Bang theory such as the horizon problem. Inflation was driven by the repulsive gravity of the vacuum, which was in an unusual state, dubbed the false vacuum.

Information, irreducible Information that cannot be compressed, or reduced, into a more compact form. The ultimate example is the number Omega, which cannot be summarised by any number shorter in length than itself.

Information, reducible Information that can be compressed, or reduced, into a more compact form.

Information Gathering and Utilising System (IGUS) An abstract entity invented by the Nobel Prize-winner Murray Gell Mann. It consists of an input register, memory registers, a computer, and so on. The way a human being experiences reality can be modelled by an IGUS. But the entity is general enough that it can even model something like a Galaxy-spanning civilisation.

Interference The ability of two waves passing through each other to mingle, reinforcing where their peaks coincide and cancelling where the peaks of one coincide with the troughs of another.

Interference pattern Pattern of light and dark stripes which appears on a screen illuminated by light from two sources. The pattern is due to the light from the two sources reinforcing at some places on the screen and cancelling at other places.

Interstellar space The space between the stars.

Ion An atom or molecule either bereft of its full complement of electrons or with more than its full complement of electrons. Unlike an atom or a molecule, it has a net electric charge.

Ionisation The process by which one or more electrons is knocked from an atom to leave an ion.

Large Hadron Collider (LHC) A giant particle accelerator being built at CERN and due for completion in 2007.

Laws of physics The edicts which orchestrate the behaviour of our Universe.

Laws of physics, fine-tuning of The observation that the laws of physics are 'just right' to permit the existence of stars, planets and life. If the force of gravity, for instance, were even a few per cent weaker or stronger than we find it, human beings would never have arisen.

Lepton Umbrella term for a group of subatomic particles including the electron and neutrino. Leptons, along with quarks, are currently thought to be the ultimate building blocks of nature. There are 6 different quarks and 6 different leptons.

Liar's Paradox The assertion by someone that: 'I am a liar.' If the statement is true, it is false; and if it is false, it is true.

Light, constancy of speed of The peculiarity that in our Universe the speed of light in empty space is always the same, irrespective of the speed of the source of light or of anyone observing the light. This is one of two cornerstones of Einstein's special theory of relativity, the other being the principle of relativity.

Light curve The variation with time of the light coming from a celestial object such as a star or supernova.

Light, speed of The cosmic speed limit – 300,000 kilometres per second.

Light year Convenient unit for expressing the distances in the Universe. It is simply the distance light travels in one year, which is 9.46 trillion kilometres.

Local group The small cluster of galaxies of which our Milky Way and the Andromeda Galaxy are the two biggest members.

Lorentz Contraction The contraction of a body moving relative to an 'observer'. The observer sees the body shrink in the direction of its motion. The effect is noticeable only when the body is moving close to the speed of light with respect to the observer.

Mach's principle The idea of the nineteenth-century philosopher Ernst Mach that bodies have inertia – that is, a resistance to any change in their motion – because of the combined gravitational pull of all the stars and galaxies in the Universe. In other words, your fridge is difficult to budge because, when you budge it, the whole

Universe pulls against you.

Magnetic Field The field of force which surrounds a magnet.

Many Worlds interpretation of quantum theory See Quantum theory, Many Worlds interpretation of.

Mass A measure of the amount of matter in a body. Mass is the most concentrated form of energy. A single gram contains the same amount of energy as about 100 tonnes of dynamite.

Mass, gravitational The mass that quantifies the response of a body to the force of gravity.

Mass, inertial The mass which quantifies the resistance of a body to changes in its motion.

Mass, rest The mass which quantifies the amount of energy a non-moving body contains.

Mass–energy The energy a body possesses by virtue of its mass. This is given by the most famous equation in all of physics – $E = mc^2$, where E is energy, m is mass and c is the speed of light.

Matter The most concentrated form of energy.

Matter, dark Matter which gives out no discernible light and whose existence is inferred from the gravitational pull it exerts on visible matter such as stars and galaxies. The Universe's dark matter outweighs its normal matter by a factor of between 6 and 7. It may consist of hitherto unknown subatomic particles.

Maxwell's equations of electromagnetism The handful of elegant equations, written down by James Clerk Maxwell in 1873, which neatly summarise all electrical and magnetic phenomena. The equations reveal that light is an electromagnetic wave.

Microwave A type of electromagnetic wave with a wavelength in the range of centimetres to tens of centimetres.

Milky Way Our Galaxy.

Modulation The continuous variation of the amplitude or frequency of a carrier wave. It is by means of such modulation that information – for instance, a radio programme – is impressed on an electromagnetic wave.

Molecule Collection of atoms glued together by electromagnetic forces. One atom, carbon, can link with itself and other atoms to make a huge number of molecules. For this reason, chemists divide molecules into 'organic' – those based on carbon – and 'inorganic' – the rest.

Momentum A measure of how much effort is required to stop a body. For instance, an oil tanker, even though it may be going at only a few kilometres an hour, has far more momentum than a Formula 1 racing car going at 200 kilometres per hour.

Momentum, conservation of Principle that momentum can never be created or destroyed.

Multiverse Hypothetical enlargement of the cosmos in which our Universe turns out to be one among an enormous number of separate and distinct universes. Most universes are dead and uninteresting. Only in a tiny subset do the laws of physics promote the emergence of stars and planets and life.

Natural Selection The idea that the traits of creatures who compete for scarce resources and survive to produce offspring are the traits that end up in a population.

Neutrino Neutral subatomic particle with a very small mass that travels very close to the speed of light. Neutrinos hardly ever interact with matter. However, when created in huge numbers they can blow a star apart as in a supernova.

Neutron One of the two main building blocks of the atomic nucleus at the centre of atoms. Neutrons have essentially the same mass as protons but carry no electrical charge. They are unstable outside of a nucleus and disintegrate in about ten minutes.

Newton's first law A body remains in a state of rest or uniform motion in a straight line unless acted upon by an external force.

Newton's second law The force on a body is the rate of change of its momentum. Conventionally, the law is written as $F = ma$, where F is the force experienced by a body of mass, m, and a is the acceleration that results. In fact, the law is no more than a definition of inertial mass, which is defined as the ratio of the force applied to a body to the acceleration produced.

Newton's universal law of gravity The idea that all bodies pull on each other across space with a force which depends on the product of their individual masses the inverse square of their distance apart. In other words, if the distance between the bodies is doubled, the force becomes four times weaker; if it is tripled, nine times weaker; and so on. Newton's theory of gravity is perfectly good for everyday applications but turns out to be an approximation. Einstein provided an improvement in the general theory of relativity.

Nuclear energy The excess energy – binding energy – released when one atomic nucleus changes into another atomic nucleus.

Nuclear fusion The welding together of two light nuclei to make a heavier nucleus, a process which results in the liberation of nuclear binding energy. The most important fusion process for human beings is the gluing together of hydrogen nuclei to make helium in the core of the Sun since its by-product is sunlight.

Nuclear reaction Any process which converts one type of atomic nucleus into another type of atomic nucleus.

Nucleon Umbrella term used for protons and neutrons, the two building blocks of the atomic nucleus.

Nucleus See Atomic nucleus.

Ockham's Razor The rule of thumb, promulgated by the fourteenth-century Franciscan monk, William of Ockham, that, if there are two competing theories both of which explain some phenomenon, the one that makes the least assumptions is invariably the true one.

Omega The jewel in the crown of Algorithmic Information Theory. Omega is a number that cannot be generated by a computer program shorter than itself. It is irreducible, incompressible information. Omega is also related to the halting problem.

Omega Point The endpoint of the Omega Point Universe. Here all light rays from the past Universe converge and an infinite amount of information processing may be carried out.

Omega Point Universe A universe which contracts faster in one direction than all other directions. In such a universe, the temperature differences grow without limit, enabling an infinite amount of information processing before the universe ends in the Omega Point.

Particle, fundamental One of the basic building blocks of all matter. Currently, physicists believe there are 6 different quarks and 6 different leptons, making a total of 12 truly fundamental particles. The hope is that the quarks will turn out to be merely different faces of the leptons.

Particle, Higgs A localised knot in the hypothetical Higgs field. There may be more than one Higgs particle. It all depends on how nature implements the Higgs mechanism for bestowing mass on matter.

Particle, virtual A subatomic particle which is permitted by the Heisenberg uncertainty principle to pop into existence as long as it pops back out again in a very short time.

Particle accelerator Giant machine, often in the shape of a circular race track, in which subatomic particles are accelerated to high speed and smashed into each other. In such collisions, the energy of motion of the particles becomes available to create new particles.

Particle physics The quest to discover the fundamental building block and the fundamental forces of nature.

Photon Particle of light and force-carrier of the electromagnetic force.

Physics, classical Non-quantum physics. Classical physics is a recipe for predicting the future – for instance, the location of Mars the day after tomorrow – with 100 per cent certainty.

Physics, laws of The fundamental laws which orchestrate the behaviour of the Universe.

Physics, quantum Essentially, the physics of the atomic and subatomic realm (although it can have manifestations in the large-scale world). Whereas classical physics predicts the future with 100 per cent certainty, quantum physics predicts only the probabilities of events – for instance, the chance that a particular atom will decay some time in the next hour. Fortunately, the unpredictability is predictable – or we would live in a world of utter chaos!

Pi (π) The ratio of the circumference to the diameter of a circle.

Planck energy The super-high energy at which gravity becomes comparable in strength to the other fundamental forces of nature.

Planet A small sphere-shaped body orbiting a star. A planet does not produce its own light but shines by reflecting the light of its star. There is currently some dispute over what constitutes a planet and whether Pluto, for instance, deserves the name.

Positron Antiparticle of the electron.

Pressure The force per unit area exerted on a container – for instance, by air molecules drumming on the inside of a balloon.

Pressure, negative The opposite of normal, positive pressure. Whereas stuff with positive pressure wants to expand, like the air in a balloon, stuff with negative pressure wants to shrink. If it were possible to fill a balloon with it, the fabric of the balloon would be sucked inwards rather than blown outwards.

Principle of computational equivalence The idea, proposed by Stephen Wolfram, that systems of equivalent complexity are truly equivalent. In other words, a system such as the Earth's atmospheric circulation, which is as complex as a living thing, has just as much right to be classed as a living thing. The principle implies that life can be implemented in a multitude of systems not just the system of biochemicals of terrestrial life.

Principle of equivalence The idea that gravity and acceleration are indistinguishable, at least in a small enough region of space. This observation was the cornerstone of Einstein's general theory of relativity.

Probability A number between 0 and 1, with 0 corresponding to a 0 per cent chance and 1 corresponding to a 100 per cent chance.

Protein A long-chain molecule composed of amino acids and used by living things for a multitude of tasks, ranging from providing cellular scaffolding to speeding up chemical reactions.

Proton One of the two main building blocks of the nucleus. Protons carry a positive electrical charge, equal and opposite to that of electrons.

Quantum The smallest chunk into which something can be divided. Photons, for instance, are quanta of the electromagnetic field.

Quantum chromodynamics The quantum theory of the strong nuclear force. The force arises from the exchange of gluons.

Quantum computer A machine that exploits the fact that quantum systems such as atoms can be in many different states at once to carry out many calculations at once. The best quantum computers can manipulate only a handful of binary digits, or bits, but in principle such computers could massively outperform conventional computers.

Quantum field theory Theory in which the fundamental entities of nature are deemed to be fields and the particles are mere 'excitations' of the fields.

Quantum fluctuations The appearance of energy out of the vacuum as permitted by the Heisenberg uncertainty principle. Usually, the energy is in the form of virtual particles.

Quantum interference The interference between the components of the quantum wave representing a particle. This is essentially the source of all weird quantum behaviour.

Quantum probability The chance, or probability, of a microscopic event. Although nature prohibits us from knowing things with certainty, it nevertheless permits us to know the probabilities with certainty!

Quantum superposition A situation in which a quantum object such as an atom is in more than one state at a time. It might, for instance, be in many places simultaneously. It is the interaction, or 'interference', between the individual states in the superposition which is the basis of all quantum weirdness. Decoherence prevents such interaction and therefore destroys quantum behaviour.

Quantum theory See Physics, quantum.

Quantum theory, Copenhagen interpretation of The idea that a particle exists as a weird superposition of possibilities until the moment it is 'observed'. It is the act of observation – ill-defined in the interpretation – that suppresses weird quantum behaviour and makes a particle behave itself.

Quantum theory, Many Worlds interpretation of The idea that all possibilities encapsulated in the 'wave function' that describes a microscopic particle are real. In other words, if a particle can be in two places at once and we make an observation that pins it down to one place, there is another reality, or Universe, containing another version of us, who observes the particle to be in the other place. In this way, events at the quantum level cause the universe to continually split into a multitude of parallel realities.

Quantum vacuum The quantum picture of empty space. Far from empty, it seethes with ultra-short-lived microscopic particles which are permitted to blink into existence and blink out again by the Heisenberg uncertainty principle.

Quark Particle out of which the protons and neutrons of atoms are made. Quarks, along with leptons, are currently thought to be the ultimate building blocks of nature. There are 6 different quarks and 6 different leptons.

Quasar A galaxy which derives most of its energy from matter heated to millions of degrees as it swirls into a central giant black hole. Quasars can generate as much light as a hundred normal galaxies from a volume smaller than the Solar System, making them the most powerful objects in the Universe.

Qubit A quantum bit, or binary digit. Whereas a normal bit can only represent a '0' or a '1', a qubit can exist in a superposition of the two states, representing a '0' and a '1' simultaneously. Because strings of qubits can represent a large number of numbers simultaneously, they can be used to do a large number of calculations simultaneously.

Radiation Energy which is radiated through space in the form of photons.

Radio wave A type of electromagnetic wave with a long wavelength, longer than about a centimetre.

Red giant A star which has exhausted the energy-generating hydrogen fuel in its core. Paradoxically, the shrinkage of the core – which is deficient in heat to hold it up against gravity – heats up the interior of the star. Furious burning of hydrogen in a ring of fire around the core causes the outer envelope of the star to balloon up and cool to a dull red colour. A red giant – the future of the Sun – often pumps out about 10,000 times as much heat as the Sun, principally because of its enormous surface area.

Red shift The loss of energy of light caused by the expansion of the Universe. The effect can be visualised by drawing a wiggly light wave on a balloon and inflating it. The wave becomes stretched out. Since red light has a longer wavelength than blue light, astronomers talk of the cosmological red shift. (A red shift can also be caused by the Doppler effect, when a body emitting light is flying away from us. And it can be caused when light loses energy climbing out of the strong gravity of a compact star such a white dwarf. This is known as a gravitational red shift.)

Refractive index The ratio of the speed of light in empty space to the speed of light in a medium. In practice, it causes light to be bent as it traverses a medium such as a piece of glass.

Relativity, general theory of Einstein's generalisation of his special theory of relativity. The theory relates what one person sees when they look at another person accelerating relative to them. Because acceleration and gravity are indistinguishable – the principle of equivalence – general relativity is also a theory of gravity. General relativity shows gravity to be nothing more than the warpage of space-time. The theory incorporates several ideas that were not incorporated in Newton's theory of gravity. One is that nothing, not even gravity, can travel faster than light. Another is that all forms of energy have mass and so are sources of gravity. Among other things, the theory predicts black holes, the expanding Universe and that gravity should bend the path of light.

Relativity, principle of The observation that all the laws of physics are the same for observers moving at constant speed with respect to each other.

Relativity, special theory of Einstein's theory which relates what one person sees when they look at another person moving at constant speed relative to them. It reveals, among other things, that the moving person appears to shrink in the direction of their motion while their time slows down, effects which become ever more marked as they approach the speed of light.

Schrödinger equation Equation which governs the way in which the quantum wave describing, say, a particle, changes with time.

Science, equation-based Conventional science.

Science, program-based The 'new kind of science' proposed by Stephen Wolfram.

Scientific notation Shorthand for writing big numbers. For instance, 10^{99} equals 1 followed by 99 zeroes.

Search for extraterrestrial intelligence (SETI) The use of telescopes to scan the sky for either a radio or optical signal from an extraterrestrial civilisation.

SETI See Search for extraterrestrial intelligence.

Set theory A branch of mathematics regarded as simpler and more fundamental than most others. A set is merely a group of entities. For instance, there is the set of all mammals, and the set of all countries beginning with 'B'.

Singularity Location where the fabric of space-time ruptures and therefore cannot be understood by Einstein's theory of gravity, the general theory of relativity. There was a singularity at the beginning of the Universe. There is also one in the centre of every black hole.

Software The infinitely rewritable instructions of a computer which enable it to simulate any other machine.

Solar System The Sun and its family of planets, moons, comets and other assorted rubble.

Space network A network of nodes connected together that may underpin space itself.

Space-time In the general theory of relativity, space and time are seen to be essentially the same thing. They are therefore treated as a single entity – space-time. It is the warpage of space-time that turns out to be gravity.

Spectrum The separation of light into its constituent 'rainbow' colours.

Square kilometre array (SKA) A giant array of telescopes with 100 times the collecting area of any existing telescope, which is being planned for operation in 2020. It will consist of a cluster of dishes within five kilometres of one another, with outriders up to 3,000 kilometres away. The central dishes could be built either in Argentina, Australia, South Africa or China.

Standard model of particle physics The picture in which the fundamental forces are explained as the result of the exchange of force-carrying particles – photons, in the case of the electromagnetic force; vector bosons in the case of the weak nuclear force; and gluons in the case of the strong nuclear force.

Star A giant ball of gas which replenishes the heat it loses to space by means of nuclear energy generated in its core.

Stochastic electrodynamics Theory in which the quantum vacuum is absolutely central to the creation of the world. Ultimately, all bizarre quantum behaviour of microscopic particles can be traced back to the relentless buffeting they receive from the ceaselessly churning quantum vacuum.

String theory See Superstring theory.

Strong nuclear force The powerful short-range force between quarks. It holds protons and neutrons together in an atomic nucleus.

Subatomic particle A particle smaller than an atom such as an electron or a neutron.

Sun The nearest star.

Supernova A cataclysmic explosion of a star. A supernova may, for a short time, outshine an entire galaxy of 100 billion ordinary stars. It is thought to leave behind a highly compressed neutron star.

Supernova, Type Ia The explosion of a white dwarf star triggered by the dumping of matter on it from a companion star. Since all such supernovae arise from essentially the same type of star, they are considered to be of equivalent intrinsic brightness. This makes them useful as cosmological distance indicators since we can be sure that a Type Ia that is fainter than another is also farther away.

Superstring theory Theory which postulates that the fundamental ingredients of the Universe are tiny strings of matter. The strings vibrate in a space-time of ten dimensions. The great pay-off of this idea is that it may be able to unite, or 'unify', quantum theory and the general theory of relativity.

Symmetry Symmetry concerns the things that do not change when an object is transformed in some way. For instance, a face that looks the same when it is reflected in a mirror is said to show 'mirror symmetry'.

Symmetry, broken A symmetry which is no longer apparent. Often the laws of physics are symmetric whereas their consequences are not. Take a pencil perfectly balanced on its point. The law of gravity is perfectly symmetric. There is a vertical force – down towards the centre of the Earth – but there is no sideways force. The sideways force is therefore the same in all directions – that is, symmetric with respect to orientation in space. But, when the pencil falls, it will fall in one direction in particular. The symmetry will be broken.

Temperature The degree of hotness of a body. It is related to the energy of motion of the particles that compose it.

Theorem A mathematical statement deduced from a set of axioms by the application of logic.

Theory of everything A theory that physicists dream of which captures all the fundamental features of reality in one simple set of equations.

Thermal radiation Radiation given out by a body that absorbs all the heat that falls on it. The heat is shared among the atoms in such a way that the heat radiation given out takes no account of what the body is made of but depends solely on its temperature and has a characteristic and easily recognisable form. Often such a body is known as a 'black body'. An approximate example is the Sun.

Thermodynamics The theory of heat.

Time, flow of A popular idea which cannot be correct since anything that flows changes with time. How can time change with time? It cannot.

Time dilation The slowing down of time for someone moving close to the speed of light or experiencing strong gravity.

Transit The passage of one celestial body in front of another as seen from our vantage point on Earth. In the case of a planet orbiting a nearby star this causes a noticeable dimming of the star's light as part of its luminous disc is obscured.

Turing machine, universal A theoretical machine devised by the mathematician Alan Turing in the 1930s and capable of simulating any other conceivable machine. In today's jargon, it is simply a computer.

Uncertainty principle See Heisenberg uncertainty principle.

Uncomputability The idea that some things in mathematics cannot be computed by any conceivable computer.

Undecidability The idea that some theorems in mathematics are true but cannot be proved to be true. This is also called incompleteness.

Unification The idea that at extremely high energy, the four fundamental forces of nature were one, united in a single theoretical framework.

Universal constructor A machine capable of constructing any other machine.

Universe All there is. This is a flexible term, which once was used for what we now call the Solar System. Later, it was used for what we call the Milky Way. Now it is used for the sum total of all the galaxies, of which there appear to be about ten billion within the observable Universe.

Universe, bouncing The idea that our Big Bang was triggered after a previous contracting phase in which the Universe shrunk down to a Big Crunch and bounced into a new expansion phase. Big Bangs and Big Crunches therefore alternate throughout eternity. The idea has several fatal flaws and so is no longer considered a viable possibility.

Universe, closed A universe that contains sufficient matter that the combined gravity of all that matter eventually brakes and reverses its expansion, shrinking Creation down to a Big Crunch.

Universe, computational The abstract universe of all conceivable computer programs. Since we can be considered computer programs, we – or at least, cyber versions of us – exist in the computational universe.

Universe, cyclic An idea in which our Universe is a four-brane – a four-dimensional island in the ten-dimensional space of string theory. Periodic collisions between our brane and another brane drive periodic bursts of expansion within our Universe. Instead of Big Bangs alternating with Big Crunches, as in the bouncing universe, Big Bangs are followed by Big Bangs.

Universe, ekpyrotic Universe in which the Universe is a four-brane –

a four-dimensional island in the ten-dimensional space of string theory – and the Big Bang was triggered by a collision between our brane and another brane.

Universe, expansion of The fleeing of the galaxies from each other in the aftermath of the Big Bang.

Universe, observable All we can see out to the Universe's horizon. The Universe has a horizon because it was born only 13.7 billion years ago. This means that we can see only the stars and galaxies whose light has taken less than 13.7 billion years to reach us. All other objects are currently beyond the horizon of the observable Universe.

Universe, open A universe that contains insufficient matter for the combined gravity of all that matter ever to reverse its expansion. Such a universe will expand for ever.

Universe, oscillating Another name for bouncing universe.

Vacuum, false An unusual state of the vacuum with sufficient negative pressure that it generates repulsive gravity. Such a state is believed to have existed at the beginning of the Universe and to have driven the super-fast expansion of 'inflation'.

Vacuum, polarised The distortion of the quantum vacuum by the presence of the electric charges in matter. If the charges in matter are positive, for instance, they will attract the vacuum's negative virtual particles and repel its positive virtual particles.

Vacuum, quantum See Quantum vacuum.

Vector boson Particles whose exchange by other particles gives rise to the weak nuclear force.

Virtual particle Particle which has a fleeting existence, popping into being and popping out again according to the constraint imposed by the Heisenberg uncertainty principle.

Von Neumann probe, self-reproducing A cross between a starship and an intelligent factory. Such a probe would reach a target planetary system and use the resources there to make two copies of itself. In this way such probes could visit every star in the Galaxy in a relatively short time.

Wavelength The distance for a wave to go through a complete oscillation cycle.

Wave function A mathematical entity that contains all that is knowable about a quantum object such as an atom. The wave function changes in time according to the Schrödinger equation.

Wave-particle duality The ability of a subatomic particle to behave as a localised billiard-ball-like particle or a spread-out wave.

Weak nuclear force The second force experienced by protons and neutrons in an atomic nucleus, the other being the strong nuclear force. The weak nuclear force can convert a neutron into a proton and so is involved in beta decay.

Weight The force of gravity acting on a body.

Zero-point fluctuations See Quantum fluctuations.

Zitterbewegung Trembling motion. The idea, proposed by Louis de Broglie and Erwin Schrödinger, that an electron is a point-like charge which jitters about randomly within a sphere of diameter the Compton wavelength.

Further Reading

Chapter 1: Elvis Lives

Garriga, Jaume and Vilenkin, Alexander, 'Many Worlds in One', *Physical Review*, vol. D 64, no. 043511, 26 July 2001. Also available at: http://xxx.lanl.gov/abs/gr-qc/0102010

Knobe, Joshua, Olum, Ken and Vilenkin, Alexander, 'Philosophical Implications of Inflationary Cosmology', *British Journal for the Philosophy of Science*, vol. 57, no. 1, March 2006: http://xxx.lanl.gov/abs/physics/0302071

Linde, Andrei, 'The Self-Reproducing Inflationary Universe', *Scientific American*, November 1994

Tegmark, Max, 'Parallel Universes', *Scientific American*, May 2003, p. 31.

Chapter 2: Cosmic Computer

Wolfram, Stephen, *A New Kind of Science*, Wolfram Media Incorporated, 2002. Also available online at: http://www.wolframscience.com/nksonline/toc.html

For more information about Stephen Wolfram, Wolfram Research and *A New Kind of Science* go to:

http://www.stephenwolfram.com
http://www.wolfram.com
http://www.wolframscience.com

Chapter 3: Yoyo Universe

Khoury, Justin, Ovrut, Burt, Steinhardt, Paul and Turok, Neil, 'The Ekpyrotic Universe: Colliding Branes and the Origin of the Hot Big Bang', *Physical Review*, vol. D 64, no. 123522, 2001:

http://xxx.lanl.gov/abs/hep-th/0103239

Khoury, Justin, Ovrut, Burt, Sieberg, Steinhardt, Paul and Turok, Neil, 'From Big Crunch to Big Bang', *Physical Review*, vol. D 65, no. 086007, April 2002:

http://xxx.lanl.gov/abs/hep-th/0108187

See also:

http://www.feynman.princeton.edu/~steinh

Chapter 4: Keeping It Real

Schulman, Lawrence S., *Time's Arrows and Quantum Measurement*, Cambridge University Press, 1997.

Chapter 5: No Time Like the Present

Hartle, James, 'The Physics of "Now"', *American Journal of Physics*, vol. 73, p. 101, 2005. Also available at:

http://xxx.lanl.gov/abs/gr-qc/0403001

Chapter 6: God's Number

Chaitin, Gregory, *Conversations with a Mathematician*, Springer-Verlag, 2002.

—, *Exploring Randomness*, Springer-Verlag, 2001.

—, *The Limits of Mathematics*, Springer-Verlag, 1998.

—, *The Unknowable*, Springer-Verlag, 1999.

Gregory Chaitin's websites (mirror images) are at:
  http://www.umcs.maine.edu/~chaitin and
  http://www.cs.auckland.ac.nz/CDMTCS/chaitin

Chapter 7: Patterns in the Void

Law, Stephen, *The Philosophy Gym*, Headline, 2003.

Stenger, Victor, *The Comprehensible Cosmos: Where Do the Laws of Physics Come From?*, Prometheus Books, 2006:
  http://www.colorado.edu/philosophy/vstenger

—, *Has Science Found God?*, Prometheus Books, 2003.

—, *Timeless Reality: Symmetry, Simplicity, and Multiple Universes*, Prometheus Books, 2000.

Wilczek, Frank, 'The Cosmic Asymmetry between Matter and Antimatter', *Scientific American*, vol. 243, no. 6, December 1980, p. 82.

Zee, A., *Fearful Symmetry*, Princeton University Press, 1999.

Chapter 8: Mass Medium

See:
  http:// www.calphysics.org

Chapter 9: An Alien at My Table

See:
  http://www.wolframscience.com/nksonline/section-12.10
  http://www.wolframscience.com/nksonline/notes-section-12.10-text
  http://www.wolfram.com
  http://www.wolframscience.com

Chapter 10: The Billboard in the Sky

Hsu, Stephen and Zee, Anthony, 'Message in the Sky'. See:
http://arxiv.org/abs/physics/0510102

Chapter 11: The Never-ending Days of Being Dead

Tipler, Frank, *The Physics of Immortality*, Macmillan, 1994.
Webb, Stephen, *If the Universe is Teeming with Aliens – Where is Everybody? Fifty Solutions to the Fermi Paradox and the Problem of Extraterrestrial Life*, Copernicus, 2002.

Index